Don,

Hope you enjoy the book.

Gene Davenport

LT COL, USAF (RET)

RADAR NAVIGATOR

SWOON 52

Enjoy these other books in the series:

We Were Crewdogs I – The B-52 Collection

We Were Crewdogs II– More B-52 Crewdog Tales

We Were Crewdogs III – Peace Is Our Profession

We Were Crewdogs IV – We Had To Be Tough

We Were Crewdogs V – We Flew The Heavies

We Were Crewdogs – The Vietnam Collection

Printed versions available by mail from
www.wewerecrewdogs.com

Tommy Towery
5709 Pecan Trace
Memphis, TN 38135
tommytowery@gmail.com

eBook versions available from
Amazon.com

ISBN: 978-0-578-10397-6

Freedom Is Not Free

We Were Crewdogs VI

Edited by

Tommy Towery

Contributing Authors

John R. Bair
Former Captain, USAF

Clif Banner
Lt Col, USAF (Ret.)

James E. Bradley
Lt Colonel, USAF (Ret.)

Kenneth G. Brown
Major, USAF (Ret.)

William J. Bushman,
Lt Colonel, USAF (Ret.)

Gene Cassanova
Major, USAF (Ret.)

Allan F. Cobb
Lt Colonel, USAF (Ret.)

Doug Cooper
Lt Colonel, USAF (Ret.)

Gene Daspit
Lt Colonel, USAF (Ret.)

Derek H. "Detch" Detjen
Major, USAF (Ret.)

Richard L. Gaines
Colonel, USAF (Ret.)

Russell Greer
MSgt, USAF (Ret.)

Jack C. Hawley
Major, USAF (Ret.)

Stephen Henley
Former Captain, USAF

Thomas D. Jones, PhD
Former Captain, USAF

Kenneth P. Katz
Former Captain, USAF

Harold "Rusty" Keller
MSgt, USAF (Ret.)

Roland R. LaFrance, Sr.
Former SSGT, USAF

CK McCusker
Former Captain, USAF

Bruno Melton
Lt Colonel, USAF (Ret.)

Gelvin Nicely
Chief MSgt, USAF (Ret.)

Vincent "Vince" H. Osborne
Colonel, USAF (Ret.)

Michael W Radke
MSgt, USAF (Ret.)

Ray Schalk
Former Captain, USAF

Kenneth Schmidt
Major, USAF (Ret.)

Dan Staley
Major, USAF (Ret.)

Robert B. Stewart Jr
Lt Colonel, USAF (Ret.)

George R. Thatcher, Ed.D
Major, USAF (Ret.)

Tommy Towery
Major, USAF (Ret.)

Dave Volker
Former Captain, USAF

Walter W. Want
Former Captain, USAF

Dave Wellman
Former Captain, USAF

Table of Contents

Dedication 1

Chapter One – Military Career 3

My Unforgettable First Year as a B-52 EW 4
A B-52 Flyer But Not A Crewdog 12
Fate, Karma, and Me 16
My Last Flight as a B-52 Crewdog 22
Some Pilots in My Career – A Reflection 25
My Decision to Leave SAC 28

Chapter Two – The Cold War 31

My First EWO Certification 32
The Night the Balloon Almost Went Up 39
Winter of 1965-66 - Glasgow AFB, MT 41
Dim Star, Bright Star—Broken Hearted Mission 43
Plotting the DEW Line 47
Cold War Duty – '66 to '68 in a B-52H 49
Friday Afternoon Follies at Carswell 51
RAF Blue Steel, A Reflection 53
Getting the Gas – B-52D Aerial Refueling 55
The B-52 and the Space Shuttle 60
What the Heck Is a MITO? 64
A Close Call for an Alert Crew 66
Close Counts in Nukes Too… 68

Chapter Three - Emergency 75

Beating the Odds - Escape from Swoon 52 76
Bird Strike at 440 Knots 89
"He" Was With Us That Morn 93

Blown Tire On Takeoff 96
First On Scene 102
A Disastrous Chrome Dome 107
Gear Head 113
A Most Tragic Weekend 116

Chapter Four–Southeast Asia 123

Arc Light and Crew Integrity: 124
Panel 2E, Row 12 128
Did You Hear That? 133
Port Bow and More 139
A Few War Stories 142
Arc Light and MARS: Staying in Touch 148
No Good Deed Goes Unpunished 152
1965 - Initial B-52F Deployment to Andersen AFB 154

Chapter Five– Southwest Asia 161

Desert Storm Combat 162
Don't Lower the Gear! 168
The "Word" 172
Baghdad by Moonlight 174
Operation Iraqi Freedom: Day One 177
Get Home-ites 181

Chapter Six – Bar Stories 191

The Kiwi Colonel 192
Ice for Christmas 196
Out of Bomber Experiences 197
Turner AFB Reprise 205
Return to Carswell 213
Trials and Tribulations 220
Welcome to Roll Call! 228
The Stag Bar – An Exposé 237
Faces 241
Always a Crewdog 242

Résumé 245

Freedom [free-duhm] - *noun* – an absence of undue restrictions and an opportunity to exercise one's rights and powers.

"Eternal vigilance is the price of liberty."
-- Wendell Phillips

"The only thing necessary for the triumph of evil is for good men to do nothing."
-- Edmund Burke

Dedication

Lothar "Nick" Maier

August 30, 1932 - December 1, 2011

This volume is dedicated to Nick, an early contributor of stories to the *We Were Crewdog* series and the author of *B*U*F*F (Big Ugly Fat F*****)*. Though we never met, we corresponded many times and shared stories and ideas about preserving the memories of BUFF crewdogs. His skills and knowledge will be missed by the B-52 community.

Chapter One – Military Career

Military Career [mil-i-ter-ee] [kuh-reer] *–noun* - An occupation or profession, especially one requiring special training, for, or pertaining to war - followed as one's lifework.

My Unforgettable First Year as a B-52 EW

Bob Stewart

The narrow winding path that led me to duty as a B-52 crewdog was one seldom traveled by other EW officers. After having spent five years teaching electronic countermeasures and electronic warfare in Air Training Command, I was hoping to transfer with the school when it was moved from Keesler AFB in Biloxi, Mississippi, and relocated at Mather AFB in California.

Such was not to be, however, and in 1962 I was selected to change places with Captain Peter Tsouprake as a crewmember on a B-52 at Glasgow AFB, Montana. Peter had originally transferred from Fairchild AFB, Washington, to Glasgow when the new bombardment wing was assigned there. Later after his Mather AFB assignment, he was awarded the Air Force Cross for his Wild Weasel heroics in Vietnam.

The news of my upcoming assignment to Montana and to B-52 crew duty did not bring joy to my heart. I was a Southern boy who did not relish spending several long years on the frigid prairies up next to the Canadian border. I had never flown in a jet aircraft of any kind and certainly never in a bomber; so, of course, I had never flown with nuclear weapons powerful enough to incinerate entire metropolitan areas. Thoughts of strapping my body into a rocket-propelled ejection seat brought on dreadful nightmares. As horrendous as these were, I soon discovered I had a lot more stuff to worry about than extreme weather, nuclear bombs, jet aircraft, and rocket ejection seats.

While at Keesler AFB my duties as both a classroom and inflight instructor were to train mostly newly commissioned college ROTC graduates for electronic warfare officer duties in various commands of the Air Force. We did not train EWs solely for Strategic Air Command. In fact, SAC EW equipment and tactics involving electronic warfare were so highly classified and guarded by SAC that we in training had very little knowledge of the B-52 and SAC operations. However, the folks at Glasgow thought since I had taught for five years already knew everything there was to know about SAC and B-52s I did not need much more training. Little did they know how unprepared I was for crewdog life.

When I arrived at Glasgow, late in 1962, I noted most of the B-52 crews and wing staff had transferred there from Fairchild AFB near Spokane, Washington. The wing needed a new crew as soon as personnel resources would allow such, so when I arrived the decision was made to form crew R-35. I would be the EW on that crew. Every one of the crew, except the Gunner and me, had previously flown in B-47s and were familiar with SAC war plans, training, testing, and crewdog life.

Our new crew was immediately sent to Castle AFB, California, for crew training. We had only been there a very short time when our training was suspended because of the Cuban Missile Crisis. All available aircraft and instructor crews were assigned ground alert duties; therefore, our crew was restricted to the base and left alone to play tennis and watch television until the crisis was resolved. We were short changed on our training because once the curfew was lifted, the flight schedules were condensed to crank out more flight crews. We were soon on our way to Walker AFB in New Mexico to complete our training.

One of the first things I learned at Walker was that EWs were expected to use the UHF radio to contact Air Defense radar sites to obtain permission and clearance to practice jamming their radars. This chore was always handled by the pilots of the TC-54D training aircraft at Keesler, so prior to our arrival at Walker, I had never used a radio. I very soon learned I suffered from "Mike Fright!" I suppose I was overly concerned I might transmit something that would be considered classified or inappropriate, and for some reason my voice would quiver like an adolescent when I tried to talk on the radio. I usually became tongue tied!

The 322nd Bomb Squadron's newest crew, R-35, was the last B-52 crew to complete the quarterly "Bar None" exercise. The crew, just formed last month, proved reliable in all phases of the training mission. Standing, from left, are: Capt Robert Stewart, Jr., electronics warfare officer; Ssgt John Bennett, tail gunner; and Maj William Clark (pointing to map), aircraft commander. Seated, from left, are: Maj Edward H. Woolsey, navigator; Maj Allen P. Gyving, radar-navigator; and 1st Lt Tommie Thompson, copilot.

I made one radio communications error while at Walker AFB that my copilot has never let me live down. On our first flight involving radar jamming I failed to position the wafer switch on my communications panel from Interphone to UHF#2 when it came time for me to obtain a jamming clearance. Those clearances were very important because ground sites became quite upset if they experienced jamming from unknown sources or at inopportune times such as during heavy air traffic. So when I thought I was calling the ground site, I was actually talking on the interphone to the crew. Naturally, the entire crew was amused, and that was all the encouragement our copilot needed to have a little extra fun. Tommie Thompson disguised his voice and proceeded to ask several wayward questions about the types of runs we wanted to make. When he next asked me to whistle Dixie for verification purposes, I was chagrined to discover my mistake! I was mortally embarrassed and would have crawled under a rock if one had been close by. Fifty years later when I answer my home phone and the caller asks me to whistle Dixie, I know who is on the other end of the line without looking at Caller ID.

I must have ruined my hearing that first year on a crew. The EW was tasked at all times to monitor two UHF radios and a very noisy long range HF radio over which practice "go-codes" could be transmitted. In addition, the EW had to monitor his radar warning receivers and his other radar receivers for determining when he should warn the crew to take evasive actions and when he needed to tune his electronic jammers to counter threats. This became a wild jungle for me personally when these were combined with crew chatter on the interphone system. As a result I used maximum volume that sometimes drove my instructors and evaluators crazy until I gained more crew experience. I wear hearing aids today.

During the first few months on the crew I soon learned the real meaning of F-E-A-R. I could hear the excitement on the interphone and radio voices of the pilots and the entire crew as we roared down a runway for takeoff, and I often wondered if there was enough concrete left for our lumbering air giant. Later inflight I feared mid-air collisions while the crew was involved with joining up with a tanker aircraft and then aligning our aircraft below the KC-135 and unloading thousands of pounds of fuel. Weather, disconnects, and all sorts of bad things could have happened during that dangerous event. Frequent delays in making radio contact with tankers and avoiding unexpected weather conditions only added to the suspense of many later aerial refuelings.

Flying low level was at first a thrilling and frequently terrifying experience for me. I knew the aircraft was not designed to fly at tree-top altitudes and to roar down valleys to escape radar surveillance. A small window on the right side of the fuselage allowed me to view the outside world if I leaned far from my ejection seat, and many times I saw pine trees whizzing by ABOVE our aircraft. Had I been riding in a small plane I probably would not have been wary of the altitude, but a huge eight-engined jet bomber was a whole different ball of wax for me.

I picked up many of my worries about safety from briefings and aircraft crash reports. SAC made sure all crews were kept abreast about fatalities, crashes, and other incidents so they would not be repeated.

A few weeks after our new crew was formed, two B-52s crashed with many fatalities in separate accidents. One was from Walker with crewmembers we remembered from training, and both crashes were caused by clear air turbulence which apparently ripped off the tall vertical stabilizers. Hearing about those accidents did not set well with my wife as I had assured her the B-52 was the safest aircraft ever flown by the Air Force. It was only after stringent flight restrictions were levied that fears were somewhat reduced.

The EW ejection seat was a big source of worry and constant fear for me. I frequently wondered what would actually happen if I rotated those handles to arm the seat, squeezed the trigger to fire the rockets, and blasted through that small hole left after the hatch blew off the top of the fuselage. If those steps were successful, I wondered if I would win the battle to avoid being impaled by the vertical stabilizer sticking up from the tail. I never trusted the engineers who designed the upward ejection system, and I remained convinced throughout my entire flying career that I would never survive if I ever had to eject.

I often wondered how I would react during a real emergency since there were so many things that could go wrong on a huge bomber with eight engines operated by a crew of six men who possessed various levels of skill, motivation, and intellectual capabilities. I heard various accounts of how one copilot jumped from the window of a bomber sitting on the ground and broke his legs when he grabbed the end of the escape rope. He jumped when he should have rappelled down the rope. And a navigator accidently fired his downward ejection seat while on the runway, and he survived! I heard about a gunner who died in his compartment from lack of oxygen when his compartment door was not secured. If those things happened to others, could they happen to me or to our crew? Thankfully, none of them ever occurred with crews I flew with, but the possibilities of disasters such as those were always on my mind.

The many types of fears, worries, and concerns remained my secret. I did not want others to know how I personally felt about all the dangers of flying in B-52s. I never wanted my family to worry about them either. I was never able to entertain them about what my specific duties were as an EW and what my equipment and tactics could accomplish. I did tell my folks at home horror stories such as the family of four who froze to death during a blizzard when traveling the 19-mile road from the town of Glasgow to the base. My dad had no concept of the B-52, and he envisioned my flying as being like that in

WWII bombers. He often asked me how many times I had to bail out and if the landings were as hard as he had been led to believe.

As a senior captain incoming from an EW school, I realized right away that I was less qualified in my new duties as a crewmember than the lieutenants and young captains on the other Glasgow crews. Those guys were wise to the ways of surviving the SAC experience. I was too proud to ask the younger guys for information and advice, and as a result, I missed out on some good tips and advice which I sorely needed. They did not realize how green I was, but I did not want anyone else to know how greatly insecure I felt about my new assignment. I somehow managed to fool most and learned to survive and fake my way by studying, listening, and refraining from asking dumb questions and committing obvious errors.

I spent most of my free time trying to digest hundreds of pages of the B-52 Dash-One operating manual. I did that because I felt it might save my life some day during emergencies. SAC had a program of testing to ensure everyone knew his duties as a crewmember for both training flights and during Emergency War Operations. If anyone failed an emergency procedures (EP) test, he immediately became suspect and required additional training and re-evaluation. The wing staff did not tolerate anyone who failed to pass its testing and interrogations.

I found myself experiencing difficulties with those tests not because I did not know the subject but rather because of the poor wording on the tests. In my previous assignment, policy forbade the use in testing of phrases that included answers with the words of "always," "never," "none of the above" and "all of the above." It appeared to me the Glasgow staff delighted in violating those rules, and I could always think of a circumstance whereby the answer I marked was not the one they wanted. To survive those tests, I had to solicit advice from trusted friends concerning the answers they considered correct. It did little good to wage war with instructors and evaluators since I was new to SAC, and I did not wish to be known as a complainer or objector.

Another concern I had in my new assignment was SAC's air training program. While it was set up to provide training to ensure aircrews and support personnel could perform their assigned tasks during wartime, it also included a rating system based on competition among the bomb wings and among crews. Since all the crews were performing essentially identical duties, a system had to be devised to recognize and reward those who outperformed others. That was especially difficult to do for the EWs. For pilots it was obvious that some could fly better than others. For the navigators the bomb scoring sites could identify how accurate the bombing was, but for the gunners and EWs it was somewhat tricky to evaluate them.

Scoring the effectiveness of an EW performance against an RBS site was frequently a judgment call by the site operator. Reliable scores were based on reaction times and accuracy of jammer tuning. An unreliable EW run was treated as a failed mission with the bomber having been shot down before dropping its bombs. If an EW reacted to a threat signal a split second too early or several seconds too late, he was charged with an unreliable run. Sometimes the site would make errors in scoring. Sometimes the weather would affect the jamming effectiveness. Sometimes the aircraft equipment was marginal. Sometimes the EW was at fault. Crews were rated on how well they were ranked and thus their ratings were reflected in their effectiveness reports which were evaluated for promotion.

On one particular training mission things were going well with our crew until shortly before we were to enter the low level bombing route at an RBS site. That was when I lost all power to my radar receivers. I immediately notified the crew of my dilemma. There was no way I could determine when and where to jam. The pilot, a former B-47 pilot, asked me, the newly assigned EW, what I recommended. We had little time to contact the command post, and I could not see how terminating the mission would not be cost productive. None of us had ever heard of canceling a mission because of EW equipment failure. So I recommended we press on regardless of EW equipment problems. That was what we did. Later, at the crew debriefing I made sure that the maintenance folks were aware of the ECM equipment problems. They later confirmed equipment malfunctions as being legitimate. After the debriefing we all headed home after what we thought had been a good training mission.

Little did we know what the fall-out would be of our continuing that mission without consulting the command post and seeking guidance from the Deputy Commander for Operations (DCO). A furious DCO notified our aircraft commander to report to his office early the next morning. That was when he was informed that our actions had knocked the bomb wing from first place on the reliability rankings in Fifteenth Air Force. It seemed the whole base was soon in an uproar. Of course, our crew took most of the blame for the fiasco, and needless to say, my OER suffered accordingly. Within a few hours

Headquarters, 15AF sent out a message to all wings stating a crew elected to continue bombing runs without EW protection thus causing the wing to drop significantly in the training standings. Our new crew became the object of discussions and soon we were the most unpopular crew on the base. Later, many confided they had never faced the same problem we had encountered, and they probably would have done the same thing we did. However, a few months later SAC did revise its training rules and what our crew did on that flight became the *modus operandi*. But by that time the world had forgotten what precipitated the change, and we were never officially pardoned or forgiven.

The trials and tribulations I experienced as a new B-52 crewmember were probably not unique. But I never heard anyone relate similar fears they possessed during their first year as a crewdog. I am sure most crewmembers felt more at home in their duties than I did. EWs and gunners usually had more dead time on a routine training mission than most other crewmembers and thus had more time to worry about things that others did not have time to fret about.

I survived my early years at Glasgow. I never developed ulcers. I never had to fire that ejection seat. I never flunked a test. Later, as an instructor and evaluator, I flew with many different crews and logged over 3,000 hours in B-52Ds. I even received two Exceptionally Well Qualified (EWQs) from the lst Combat Evaluation Group and the Chief, Wing Penetration Aids. While on a select crew I was able to implement many improvements in the testing and evaluation of EW officers. Those actions led to a belated promotion and an unexpected staff position. In hindsight I concluded all my worries and fears resulted in learning experiences which were rewarded in later assignments. I learned some valuable lessons which I later applied to other phases of my personal life.

A B-52 Flyer But Not A Crewdog

Ken Katz

Previous contributors to the *We Were Crewdogs* series have been rated pilots, navigators, or gunners, and assigned to the Strategic Air Command (SAC) or its successor commands. During my Air Force active duty from 1985-89, I was a B-52 flyer - but was neither a crewdog in the traditional meaning nor assigned to SAC.

First a little background. By the mid-1970s, what was intended to be the last round of B-52 modernization (SRAM, EVS, Rivet Ace Phase VI electronic countermeasures) was winding up and B-52 flight test activities wound down in expectation that the B-52 would be replaced by the new B-1A. When President Jimmy Carter cancelled the B-1A program in 1976, the B-52 modernization was revived, with an extensive flight testing effort for the Air-Launched Cruise Missile

(ALCM) and the Offensive Avionics System (OAS) at the Air Force Flight Test Center (AFFTC) at Edwards AFB. By the time I reported in at AFFTC in 1985, both ALCM and OAS were operational, but testing was underway on the Advanced Cruise Missile (ACM), the Common Strategic Rotary Launcher (CSRL), and the OAS Block II software required for ACM and CSRL. The testing was being conducted by the Strategic Systems Combined Test Force, which consisted of AFFTC, SAC, Air Force Operational Test and Evaluation Center and contractor personnel.

As an engineer at AFFTC in the Strategic Systems Combined Test Force, my primary duty was to plan tests, analyze data, and write reports. But the real attraction of the assignment was the opportunity to fly, an uncommon thing for non-rated officers. There were a limited number of flight test engineers on flight orders, and around the time I was promoted to first lieutenant it was my turn.

Flight test engineers flying on the B-52s at Edwards were called instrumentation operators (IOs). A description of the modification to our aircraft will help explain what the IOs did.

The various test programs brought different B-52Gs and B-52Hs to Edwards - each modified and instrumented uniquely. Usually the defensive avionics and guns were either removed or disabled. The airframe and electrical system were instrumented. The OAS was a digital system and generated enormous amounts of data that was recorded on hard disk drives and tape drives. The video signals to the OAS displays were also recorded on video tape recorders. If the aircraft was going to drop weapons, then it carried cameras to record the separation dynamics. On a B-52G, the special equipment was placed wherever it could fit; on the B-52H it fit nicely in the crew bunk area. Often the EW seat was removed and replaced with a Completely Integrated Reference Instrumentation System (CIRIS) pallet. CIRIS used a network of ground-based stations in the western United States to update an onboard inertial navigation system and provided a time, position, and velocity reference against which OAS and missile navigation accuracy could be compared. The IO sat in the gunner's seat, facing some instrumentation and camera control panels, but often he was out of the seat, changing data and video tapes.

Flight tests were complicated affairs, with elaborate planning and briefings. By far the most complicated tests were live missile launches. Live launches were invariably preceded by captive carry tests. During a captive carry test, both the missile and the OAS had software loads with captive carry patches. The missile was never actually launched. Instead, it flew its mission plan under the wing of the B-52, feeding guidance cues to the OAS that drove the flight control indicator followed by the pilots. It was a good way to make sure that the missile mission plan and software worked without expending a missile.

AGM-136

On the day of the live launch, the test director and his team were located in the Ridley Mission Control Center at Edwards. The B-52 with the test missile was launched first. There were several different trajectories used for cruise missile testing. A common one was to launch over the Pacific Ocean off Point Mugu, then fly north of Santa Barbara and east to the Edwards range complex on the IR-200 route, then northeast through the Nellis range in Nevada to the Utah Test and Training Range. As you can image, the amount of coordination was staggering. At the launch point, the B-52 formed up with two F-4 missile chase aircraft and a T-38 photo chase aircraft. The F-4s had remote control equipment which the backseater could use to take over missile guidance, a necessity when flying outside of restricted airspace. One F-4, called Low Chase, followed the missile closely. The other, called High Chase, kept an eye on Low Chase and made sure that it did not fly into terrain. There was also another pair of F-4 missile chase aircraft in the operation, with one pair refueling from a KC-135 while the other pair chased. Flying further behind the missile and its chase was the EC-135 Advanced Range Instrumentation Aircraft (ARIA). ARIA had a large steerable antenna in its bulbous nose which it used to act as a flying telemetry relay and recording station, feeding data to the mission control center even if the missile was out of sight of ground stations. The aerial circus also often included helicopters to perform a mid-air retrieval of the missile at the end of its flight.

The B-52 IOs duties were essential to record the test data on the B-52. In flight, he monitored the instrumentation to verify that it was working properly and coordinated with personnel on the ground to troubleshoot and resolve problems. Throughout the flight, he changed data and video tapes. Approaching the launch point, the IO strapped into his ejection seat and prepared for his single most critical duty, which was to start the cameras at exactly the correct moment before missile launch. The aerodynamics of stores separation are difficult to analyze, and unanticipated missile motion after launch could be hazardous to both the B-52 and the missile. So it was essential to take high-speed motion pictures of the missile from several angles to compare the actual trajectory to the pre-launch calculations. The high-speed movie cameras consumed film at a ferocious rate, so they only carried a few seconds of it. If the IO initiated the cameras either too early or too late, the data would not be collected. Needless to say, nobody wanted to be involved in that type of screw-up.

There was another kind of flying that IOs did. With all the planning, aircraft modifications and ground tests, and the small number of missiles in a test program, the modified test B-52s did not do all that much flying. So SAC brought in operational B-52s once a month for the Edwards crews to maintain their proficiency. Some IOs flew on these "Rent-a-BUFF" missions just to log the hours for flight pay. But for me, it was an opportunity for hands-on learning about the aircraft that I was testing. Sitting in the IP jump seat during patterns, approaches and air refueling, acting as navigator during low level flights, and the occasional opportunity to fly the BUFF from the copilot seat were both fun and a great learning experience.

We IOs didn't sit alert and we didn't get shot at over Vietnam or Iraq. But flying over the peaceful skies of the western United States, we helped provide the weapon systems for those who did, and in our unique way contributed to the legend of the B-52.

On a personal note, a few years ago I visited my old unit, now known as the 419th Flight Test Squadron / Global Power Bomber Combined Test Force, while I was doing research for my book about the B-52. The company grade officers serving as flight test engineers and IOs were in diapers when I was on active duty and in some cases they were testing the replacements for the systems that I tested. I felt very old!

Fate, Karma, and Me

Gene Cassanova

Whack...click....whack....click....whack....click, oh shiii...iit! (Yes, in Texas that is a two syllable word.) Another victory 21-14, and another quarter in the free money jar I filled while playing ping pong at the Carswell AFB alert facility. Okay, Sgt Hauser was a pretty good player but unfortunately for him, not quite up to the level of yours truly, "Gene, Gene, the ping pong machine." Alert tour after alert tour Sgt Hauser would challenge me to several games of ping pong at $0.25 per game. I felt bad taking his money, but in reality he was just paying for fairly inexpensive lessons from a pro. But I digress. How in the heck did a young boy from Metairie, Louisiana, end up hustling ping pong on alert in Ft Worth, Texas?

As we all know, real life rarely follows the envisioned path. As chronicled here, my journey from civilianhood to Carswell was anything but simple. In 1965, the unpopular Vietnam conflict was still going strong. Being of relatively sound mind for a recent high school graduate, I realized that pursuing a college degree was not only a fiscally but also physically sound endeavor. Since I had always wanted to fly, and more importantly, didn't want to be a grunt in the jungles of Vietnam, joining the Air Force's Delayed Enlistment Program (DEP) seemed the best possible alternative to allow me to obtain a degree and at least temporarily avoid the realities of war.

Today the DEP allows individuals to join the active duty military, reserve a job, and then wait until a "slot" in basic training opens up. Individuals in the DEP are technically in the inactive reserves. They take the enlistment oath, sign a contract, and can be legally called to active duty. Back in the Vietnam era, the DEP allowed young men to complete their college education before joining the active force, maybe even as an officer and a gentleman.

Hopefully, I could remain in college, graduate, and then if the Vietnam war (or as history refers to it "a Cold-War-era military conflict") still wasn't over, I could become an Air Force flyboy for a few years prior to becoming a civilian airline pilot. That was the plan.

As I mentioned earlier, the simple path in life is typically filled with numerous twists and turns. After dutifully completing the DEP paper work process, I continued my undergraduate studies at LSU. Yes, that was in the days of Coach Charlie Mac and his penchant for three times up the middle, a cloud of dust, and then punt. A plan that was good for numerous 7-3 seasons but no national championships. But I digress again. Just before my planned graduation in January, 1970, the government instituted the Vietnam Draft Lottery.

That lottery drawing - the first since 1942 - was held on December 1, 1969, at Selective Service National Headquarters in Washington, D.C. The event determined the order of call for induction during calendar year 1970, that is, for registrants born between January 1, 1944 and December 31, 1950. Reinstitution of the lottery was a change from the oldest-first method, which had been the determining method for deciding order of call.

With radio, film, and TV coverage, the capsules were drawn from the jar, opened, and the dates inside posted in order. The first capsule - drawn by Congressman Alexander Pirine (R-NY) of the House Armed Services Committee - contained the date September 14 - so all men born on September 14 in any year between 1944 and 1950 were assigned lottery number 1. The drawing continued until all days of the year had been matched to lottery numbers.

Well, Fate, Karma, or whatever you want to call it stepped in again and thanks to my birthday (thanks mom for having me two days earlier than planned) I was the proud owner of draft number 350. Can you believe it, 350 out of 365? In other words I would only be drafted after the first 349 numbers and associated birth dates had been called. Now if I had been number four, or number 52 or even number 105, I wouldn't even have considered withdrawing from the DEP. But, wow number 350; no way in the world would I be drafted. It was as if fate had said thanks but no thanks. Anyway, after several concentrated seconds of deliberation, we decided that this was definitely a sign for me to forgo the military and continue my civilian life.

After graduating from LSU, my wonderful and loving wife Marcia and I moved back home to New Orleans hoping to continue my education or find great jobs in the crescent city. Yeah, you guessed it. Good old Karma stepped in again and helped us decide that maybe voluntarily joining the Air Force without the spectra of the draft might just be our best option. So a year after withdrawing from the DEP, I was back at the USAF recruiter's office that time seeking immediate entry to Officer Training School (OTS) and then into pilot training. We both had matured a lot in that one year and realized that becoming an Air Force officer and flyer would be a great way to start a career. Our youthful feelings of invulnerability and the glamorous vision of flying jets, seemed so exciting that the reality of the inherent dangers were never seriously considered. So in February, 1971, it was off to Officer Training School (OTS) at Lackland AFB, San Antonio, Texas.

Three short but fun-filled months later I was commissioned a second lieutenant and headed off to Undergraduate Pilot Training (UPT) at Reese AFB in Lubbock, Texas. I know I said pilot training, but what I meant to say was Pre-Undergraduate Navigator Training (UNT). Just prior to soloing the T-37 (tweet), the crack Reese AFB Instructor Pilot (IP) staff decided that my skill-set would be totally underutilized as a mere pilot and that telling pilots (and most everyone else for that matter) where to go would be the best use of my innate abilities. So after a short visit back home, Marcia, my ever-ready partner and I headed to Mather AFB in sunny Sacramento, California.

Mather is one of those kinds of assignments that is even better than you could have possibly imagined. While Sacramento itself is an extraordinary city located at the convergence of the Sacramento and American Rivers, as they say in real estate "location, location, location is everything." To young, eager lieutenants and their wives, Sacramento's location with San Francisco to the southwest, Lake

Tahoe to the east, Disneyland, Hollywood and Los Angeles to the south, to say nothing of the redwood trees and surrounding vineyards made it more of a vacation retreat than an assignment.

While UNT was certainly rigorous, there was ample opportunity to begin exploring the surrounding areas. Upon graduation from UNT, we realized that there were still many, many places in California that we wanted to visit before moving on. The best way to stay at Mather for a bit longer was to attend Navigator Bombardier Training (NBT). Unfortunately, most students graduating from NBT ended in assignment to a B-52 squadron and the not so glamorous life as a BUFF crewdog. But the smart assignment folks knew exactly how to induce gullible young lieutenants to volunteer for NBT and be happy about it. A single assignment to an FB-111 for the lucky winner of the coveted Lenzi Trophy awarded to the top simulator student from each class was more than enough to fill NBT quotas. I know it sure made my decision easy. All I had to do was work harder than any of the other new navigators (and be really, really lucky) and I was practically assured of copiloting not just directing the then hot FB-111.

Well by now, I am sure you guessed it, Karma/Fate again took over and mine was the first NBT class in several years in which there was no FB-111 slot authorized. Not to worry, I still received and actually still have the coveted Lenzi trophy proudly displayed on my bookshelf (yeah, a constant reminder of my friend Karma). So yes, now we are getting closer to how I ended up at Carswell AFB, Texas. When the NBT staff realized there were no FB-111s for my class, they panicked. How would that look to all those young men? More importantly, what would it do to future NBT quotas?

Don't listen to what anybody tells you. The Air Force is nothing if not cunning. I really think they had a plan-B even before lawyers supposedly invented that tactic. Miraculously, from out of the clear blue yonder (as the song goes) a slot to McCoy AFB in sunny Orlando, Florida, appeared on our assignment selection list. No, it wasn't there before the FB-111 glitch but suddenly it was proudly waiting to be selected. Now, what initially seemed to be an unfortunate twist of fate soon turned out to be my wife's dream assignment come true. A Mouseketeer (especially Annette Funicello) fan since early childhood, Marcia easily convinced me that being stationed in the world of Disney was far more desirable than me being a semi-copilot on a sophisticated fighter-bomber.

Well, since the Air Force had removed that option from the table anyway, I was easily swayed by my loving wife's childhood fantasy. Then Karma stepped in again. Rumors that McCoy was on the congressional chopping block and would soon be closed certainly made me reconsider my upcoming assignment selection decision. But talk about coincidence, the McCoy AFB wing commander just happened to drop in to Mather to visit with the NBT students.

Being a young and somewhat naive lieutenant, I had rarely seen, much less chatted, with many full colonels. But when I realized that my immediate future depended upon information the colonel possessed, I summoned up my courage, walked right up, and simply asked him if the rumors concerning McCoy's imminent closure were true.

"No, no, son, I can guarantee you that those stories in the press are all just that, stories. McCoy is absolutely vital to SAC and will be so for the foreseeable future. When the time comes, you select McCoy, you will be glad you did. I guarantee McCoy won't be closing anytime soon."

Based on those words and the desire for continued matrimonial bliss, we were soon headed to McCoy and the land of endless orange groves and Mouseketeers. Maybe I should have been suspicious, but being so naive I didn't correlate immediate base housing availability for a newbie second lieutenant with the inevitable. The fact that there were no B-52s at McCoy was not a real giveaway since most of the BUFFs were still deployed to Thailand. After being at McCoy for less than a week, we received TDY orders for Carswell, to complete B-52 upgrade training. For the three weeks before heading to Ft. Worth, we settled into our wonderful, spacious three-bedroom, two-bathroom base house. After setting up this wonderful home with a washer, dryer, lawnmower, curtains, furniture and carpets (white shag) we were off to Texas.

Less than halfway through the six-week upgrade training at Carswell we received the news that McCoy was, in fact, on the congressional chopping block and base closure would soon commence. It was time to scramble. Everybody currently stationed at McCoy would soon be reassigned, either to a base of their choice (assuming they could find an available slot) or one randomly selected by the bean-counters also known as Air Force Military Personnel Center (AFMPC) slugs.

Not knowing what Karma had in store me that time, I decided to take matters in my own hands and at least attempt to find a suitable substitute for McCoy. Since we really liked Carswell and the Fort Worth area, and while there was no Disneyland, at least there was Six Flags so I decided it might be a great location. Being the silver tongued devil that I envisioned myself as, I walked right into the 9th Bomb Squadron Ops Officer's office and explained that with McCoy's imminent closure, I was a free agent and would be a perfect fit for their squadron. What with me being a Lenzi Trophy winner and an excellent navigator he should snap me up before some other less deserving squadron like the 20 Bomb Squadron down the hall might do.

And that my friends, is how my almost nine-year crewdog career as a ping-pong hustler at Carswell AFB began.

My Last Flight as a B-52 Crewdog

Russell Greer

When word came down that gunners were being pulled off the BUFF effective October 1, 1991, I was assigned to the 436 STS at Carswell AFB, Texas. I was the B-52G Curriculum Development Manager for SAC and also worked with the other gunners assigned to the 436th Gunner Training Unit which was a B-52 G-to-H difference course. Some of the courses that Scott Smith (B-52H Curriculum Development Manager) and I had completed over the previous months were the B-52G & H Strange Field Disarming videos as well as the B-52G & H Aircraft Servicing courses.

As a G-model gunner at an H-model base since 1989, I had the good fortune to be schooled on the ASG-21 Fire Control System by some of SACs finest: Harley Gomes, John Wing, Pete Gertz, Chris Austin, and Scott Smith. That led to dual qualification and I flew monthly with either 9th or 20th BS crews. I also traveled to the various

G-units to maintain currency on the ASG-15 as well as roll out new training programs and review the various courses that we offered.

About two weeks prior to the announcement that CINCSAC was pulling gunners off crews as a cost-saving measure many of us throughout the command had heard the rumors that the end was near. We had already begun to get our ducks in a row, as far as making career plans for our lives after the BUFF. I had decided that being a flight engineer was the way I wanted to head if given the chance, but first I had to wait for the hammer to fall. Once the death-bell sounded we were given two weeks to make our choices and apply for retraining, retirement, or separation. The majority of us at the 436th were at CBPO the next day getting the paperwork started. Scott, Chris, and I applied for engineer slots with John deciding on Bomb Nav Technician as I recall. Harley decided to separate and I cannot remember what Pete decided to go into.

With a few weeks to go until October 1st we busied ourselves beginning to shut down the Gunner Training Unit (GTU) which involved such things as disposing of the simulator and the M-61 20mm Gatling Gun trainer. We received permission to donate the M-61 to the Armaments Museum at Eglin AFB and as of a couple of years ago it still sits proudly on display bearing a plaque with the inscription that it was donated by the gunners of the GTU with our names on it.

With just a couple of days left I attended my last mission planning session as a B-52 Gunner with a crew from the 9th BS. It was planned to be a standard sortie with low level at IR-276, A/R, a nav leg and about an hour of crash and dash (approaches and touch-and-gos) back at home plate. The morning of the final flight began as so many others had: meet at base ops, pick-up the lunches and coffee, attend the weather briefing, and then head out to the jet for preflight. It began normally enough but during the preflight it kept occurring to me that this was it - the last time I would probably ever strap into the BUFF as a member of a SAC crew and yes, it was a sad time for me.

Takeoff and A/R went as briefed but on the way to the low level route the OAS began to have numerous malfunctions and unfortunately resulted in the aircraft commander deciding there would be no low-level that day. After consulting with the command post we headed back to Carswell and ended our day with only two approaches and then a full stop. After years of eight hour and longer sorties my last clocked in at just 4.6 hours. I have a photo of that last flight on B-52H 0017 that I look back on with fond memories. I ended up with far more hours in the G, but in the end the H was what I rode into the sunset. It really seems hard to look back now at the end of 2011 and realize that 20 years have come and gone since that last flight. Where did the time go?

Scott and I ended up staying at Carswell until the end of January, 1992, before we headed off to Altus AFB, Oklahoma, and flight engineer school together. We both went to C-141s - he to Charleston AFB and me to McGuire AFB. I had some great adventures as an engineer but by far the best times I had in my entire Air Force career was as a gunner in SAC.

It is true what we always said, "Nutin' could be funner than being a gunner."

Some Pilots in My Career – A Reflection

Mike Radke

I don't recall much about the early pilots I had in my career as a B-52 gunner - they were somehow insignificant in my memories. Participating in Arc Light bombing missions changed that for obvious reasons. Combat is different than training.

My first Arc Light tour was with N-53, a newly formed crew from the 322BS from Glasgow AFB. Chet, the pilot, was a major and lst Lt Bilbrey was the crew's copilot. Chet really knew his stuff and had previously been a standboard pilot at Castle AFB. I do not recall his complete military history but I remember he only had a 10th grade education, and had been drafted. During the Korean conflict there was a shortage of fliers and he found himself in flight training and became a pilot. On our initial flight to Guam, the copilot stopped feeding fuel to the engines on our penetration and we recovered in a dive maneuver thanks to his skills.

My most memorable flying experience was something that happened after a 13-hour combat mission. On the penetration into Guam the right forward gear and the right outrigger didn't extend. The RN checked and found the cable to the banana clip had broken. Being low on fuel, a tanker was scrambled and due to some airspeed issues we ended up having to refuel in a descent. After an hour or so of trying to get the gear problem fixed we were instructed by the command post to bail out. The Navy had lost some crewmembers to sharks just the week before so there was a lot of radio communication with the controllers who threatened us with court-martials if we didn't comply. In the meantime the pilot told me to come forward, and monitor the RN in his attempt to free the latch.

The RN was Bill, a crusty gray-haired older major, with lots of experience and he had the idea to try something he had once heard about. Desperate measures called for desperate times. I was glad it was dark because I hate heights and the plan was to secure an escape rope around my waist and the nav team would support me while I climbed down and kicked the latch free once the circuit breaker had been pulled. Luck was with us that night – it worked and we were on the ground after a flight of 16 hours.

Despite the previous threats, we were not court-martialed, but instead were sent on an R&R to Japan the next day. The copilot elected to stay on Guam instead of going to Japan with the rest of us. Bilbrey was from Tennessee and an avid spelunker (cave explorer). Somehow he had discovered a cave on the island that still had munitions from WWII in it and wanted to go exploring. He messed up in his attempt and had to be rescued from the cave. The only smart thing he did was to have a partner along who went for help.

As for the rest of us, the EW and nav team stayed at Tachikawa to visit friends and Chet and I went to Tokyo via the bullet train. He signed me in as a captain at the hotel contracted for officers. After a great meal, we went over to the Nichigeki Music Hall for the show and watched two hours of small operettas - all done in the nude. It was a great experience at the time. The next couple of days we just toured, but way too soon it was back to the Rock.

Upon our return Chet made lieutenant colonel and we learned we would lose him from our crew after our tour was over. He would be going off to AFIT to finally get a college degree. The only things that happened on the rest of the tour were just the usual landing glitches - no outrigger problems or bombs resting on the bomb bay doors. Upon our return to Glasgow, we did a lot of camping, fishing, and hunting before we rotated back to Guam. At least we missed out on the Montana winters.

I had many pilots the 10 years I was assigned to KI Sawyer. Each of them had their plusses and idiosyncrasies. John was an ex-Marine grunt who took advantage of the GI Bill to get his degree. He was a really good pilot, but was still a macho Marine. In the Seventies a crew member needed the commander's okay to get a vasectomy. I can recall the papers saying it was an office procedure - walk in and walk out. I did not find that true since I was incapacitated for a couple of days when I got one. John must have believed the propaganda and decided he would have it done in town two days before a scheduled 24-hour Chrome Dome flight.

A few hours into the mission and after coming off the tanker, the pain was so unbearable he had to hit the bunk. Instead of aborting the mission, we all took shifts filling his seat for him. He could only manage a short period of seat time himself before the pain would hit him again. After landing an ambulance was dispatched to pick him up. It was a high price to pay to complete the mission - he lost one testicle.

Another pilot, Joe, was a little guy, who probably only barely met the minimum height requirements. His short coming as a pilot was taxiing, primarily because he couldn't reach the pedals. When he was taxiing into the Christmas tree while on alert, I had to put seat cushions behind his back to enable him to fully negotiate the turn. We always thought he'd break a blood vessel while doing that because his face would get bright red and the vessels would stand out.

Steiger was my S-03 pilot. He was an ex-Air America type and somewhat nuts by most people's definition. He treated me very well however, but hard-nosed the rest of the crew for some unknown reason.

A multitude of crewmembers come and go over a 20-year plus period - some memorable and some not. I was only on S-01 for a short period of time before there was a crash in a snow storm at KI Sawyer. It was on a check ride on 1 April 1977 and B-52H 60-0039 went down just short of the runway. I was DNIF for that flight. I Remember the IP Jim, copilot Chris, IRN Denny, nav Steve, and IEW Jim, all of whom I had served with on various crews over time. The AG, John, was on his solo flight returning from courier duty and had taken my place. One can never forget the personnel losses.

My Decision to Leave SAC

Walter Want

I had been selected Junior Officer of the year, served as Mister Vice for the Dining Out, and was on an early upgrade list for AC as an ICP on a select crew. As a senior first lieutenant with a very pregnant wife I was moved into base housing that was normally assigned to majors and above. There was something about those long cold winters and a need to cuddle to stay warm. We got a new car with a good heater, and a duplex with a garage. Wow! We learned that the upstairs bedroom window also made for a winter exit when the front door and garage doors were iced over and not available for access. Life was good for a young man, young wife, and a new baby.

Then it happened. The war in Vietnam escalated and crews were beginning to be moved out for 179-day TDY duty assignments in Guam and Thailand. The potential for my selection was great since we were an iron bomb qualified crew and I knew the writing was on the wall.

One day I was called into the Squadron Commander's office and was presented with a kind of an offer I could not refuse. If I accepted a regular commission, I would be upgraded and more than likely miss a SAC deployment. I was born with a big chunk of "fairness" in me and I was a bit insulted by the subtleness of the offer. I needed to make a decision. I did not want to be SAC career guy on alert. I had a college education, a young schoolteacher and a CPA-candidate professional wife, and a beautiful young daughter. SAC was not my life long career choice. I refused the offer and a very short time later was advised that I had made a bad choice and the squadron had orders for me for SEA.

I knew that was a potential option, for a serious number of the old heads were retiring and joining the airlines and perhaps just did not want the extended TDY in SEA either. Table talk on the pad was very interesting. My plight was discussed and I knew that having a brother in the Army and one in the Navy and me in the USAF my chances were better than theirs. I rolled the dice, refused the very generous offer and accepted an invitation to Forward Air Control University (FACU) in Phan Rang, Vietnam, as an O-1, O-2, or perhaps an OV-10 pilot.

I knew I had made the right choice when I was on leave in Southern Idaho visiting the wife's family and I got called back to Grand

Forks AFB for the follow-on flight during an ORI. Every aircraft was required to fly and as a recent upgraded aircraft commander I fully understood the importance of the mission.

I arrived, briefed, and was ready to go as a CP to a major that I had never met. We had the oldest A/C, 60-0024. It was the high time machine and was put back together to fly this important mission. After taxing out the Old Man came aboard and placed his hands on our shoulders. He explained that I was a known quantity and he was sure the new major was fully qualified, but if a situation arose we were to discuss it on pilot common and make the correct decision. If things got critical and we did not agree, the AC was to go with my decision. Wow! It was a rock and a hard place situation.

We took off and lost an engine on takeoff while at max gross weight. Oh what an evening it would turn out to be. We could not get up to refueling altitude so we requested a big box and stayed down as we got heavier. He was shaky at his first night refueling behind the KC-135 and needed help as we kept breaking contact. We got our load and headed out on a high altitude nav leg. Somewhere about that time the Hound Dog under the right wing went wild and a small fire broke out. We had red lights in the cockpit and we shut it down. There was no reason to report that since we seemed okay and we did not want a black mark on our ride or the new AC.

Sometime later during our high altitude bombing leg we lost an engine on the left side. We used a little rudder to kept the nose straight and again saw no reason to get excited or report it; so press on professional crewmember. During low-level entry over Louisiana we must have eaten a few birds and we lost another engine on the right side. There were no lights, just an unwinding RPM and complete loss of oil pressure. We were on time and doing as well as could be expected as a last-minute put-together SAC crew on an important mission.

We were flying low level on terrain avoidance radar, doing 425 knots and doing fine when we hit a flight of migratory birds. Yes they were birds, resulting in debris all over the windscreens and two fire lights glowing. We professionally shut two of those big hummers down and pressed on by pushing the throttles up with an AGM- 28 assist for balance on the left side.

Just after the low Pop-Up Procedure maneuver we talked quite frankly about our situation and decided to call it quits and confess our plight. I put out the radio call as a crippled nuclear-loaded SAC bomber expecting something more than a "Standby One" response. We leveled at 10,000 feet and headed north on a homeward track to Grand Forks.

The radio opened up communication with a simple question. "Do you know where you are?"

"Yes," we replied.

"Do you know who is president of the USA?"

"Yes," we replied.

"Do you know where he is from?"

"Yes," we replied and then realized we were over President Johnson's home state of Texas. We were directly asked if we could make it to Oklahoma. It meant "No Nuclear Incidents in Texas, You all hear?"

"Yes," we replied and we flew north to avoid any problems.

We later landed, with me making the landing from the right seat. He was career and I was headed to SEA as a SAC donation to the war effort in an O-2A.

I do remember as we made our landing straight-in, and that the tower controller asked if he could do anything else for us prior to the weenie roast. There must have been a thousand flashing red emergency lights along the runway and without thinking I requested they turn those XXXXXing flashing lights out because they were scaring my crew. We made a good landing and shut it down in the overrun. The only comment made as we debriefed was the use of my language regarding those XXXXing lights.

I was out of there forever. It was my last ride. Mission accomplished, but it was a bit of a game and I choose not to be a player. I recently retired and as a professional safety officer. I do know what was unsafe that evening.

I spent my SEA tour flying as a FAC from Channel 89, NKP, or Naked Phanny, where no airplane had a jet engine and many had no marks at all on them. I am a Member of the Laotians Road Patrol, and spent a few trips very North and was involved in a number of SAR missions. I have seen Hanoi from low level in a quiet airplane.

I am also very proud to say I was a SAC crewmember who was bombed twice near Mu Gia Pass by the B-52s. You Missed. Thanks!

Chapter Two – The Cold War

Cold War [kohld] [wawr] – *noun* - A state of political tension and military rivalry between nations that stops short of full-scale war, especially that which existed between the United States and Soviet Union following World War II.

My First EWO Certification

Dan Staley

On January 3rd of 1973 I reported to Grand Forks AFB as a brand new "highly qualified" B-52 copilot. As ready as I was to dazzle the hard bitten crew force of the 319th BW with my vast knowledge of aircraft systems and my highly honed aeronautical skills, all anyone could talk about was "it's true, you're a second lieutenant!" People wanted to know how old I was, gunners patted me on the head and told me that I would grow up one day (if they allowed me), and the folks in the Bomb/Nav shop made me get a note from a "grown up" just to come into their office. The Wing Commander (Col Lee R. Senter) introduced me at Officer's Call as the only second lieutenant in the Wing and the first one anyone could ever remember being assigned.

As part of my checkout to become combat ready, I not only had to demonstrate a mastery of the aircraft but also an understanding of the top secret SAC's SIOP war plan. This was done by certifying a sortie of the wing's Emergency War Order (EWO) commitment. As anyone who went through it remembers, this certification was given by the various department heads of the DO's staff (EW, Bomb/Nav, Command Post, Squadron, Intel, and Current Ops) plus select members from the wing staff to include the Wing Commander– not exactly a forgiving crowd. Just to spice things up a little, I was the only new guy in the squadron in almost a year and was certifying after the mass semi-annual crew certification cycle had been completed so I would be going through the process alone. That had the added benefit of allowing the staff to recover from the dredge of the mass certifications and be ready for new meat.

Not to be intimidated, I committed myself to putting my best effort into study and preparation. Command and Control was my first challenge, but, having mastered colors in the first grade, I quickly caught on to the concept – Green was practice, Yellow was an inconvenience, and Red, well, let's not go there. "Two Officers" verses "Two Man" was my next challenge. Discerning the difference between the two policies proved harder than one might think on the face of it. After trial and error, I finally solved the difference by resolving to always have three people around me whenever I did anything near the airplane – that is another story for another time, however. Finally, I got to the good stuff; all those red-covered books in the vault – all the volumes of SAC Tactical Doctrine.

Let's take a second and talk about the vault. I am sure the location had several names depending upon the base, but at Grand Forks it was called "The Vault" and was located in the alert facility inside "the cage." Our vault was staffed by several lieutenant colonels who, for all intents and purposes, had found a place to retire. That was fine by me; they knew their stuff, they were willing (if not eager) to help, and perhaps the biggest aid to me – they were available. As I said before, I was on my own to do my preparation, and it was very much like a blind man trying to describe an elephant, but, still, I was happy for the company. I took the guidance they gave me literally and to heart, "Don't read your presentation; Alpha hates when people read; you have to KNOW it," "know all your material verbatim, especially bold print," and "don't ever pick up the wrong manual when you are looking for a small print answer." If the intent was to motivate me by fear, it worked. However, these sound military axioms were in one way or another going to haunt me during this first certification crucible.

I worked hard on my preparation. I spent hours reading every manual and doctrine they had. I compiled lists of questions and correlated them between documents. "Why does Vol. 1 of Document X say this and Vol. III say that?" "What does this phrase mean in this circumstance, and how do I apply it in that situation?" I wrote and rewrote Tac Doctrine bold print until I could write every one of them from memory in order with page references. I memorized single-engine failure speeds and power settings, fuel logs, gear extended and AR door-open fuel degradation factors, AGM-28 launch envelope parameters, and airspeeds, to say nothing of takeoff data variables and their impacts.

I memorized bomb run times and heading, check points, altitudes, visual landmarks, and target description until I could draw the entire route of flight from memory. I reviewed and memorized radio frequencies, call signs, and recovery base procedures until I could recite them in my sleep, and probably did. I read and re-read every volume of doctrine The Vault had to offer. Finally, I wrote, rewrote, and re-rewrote my Emergency War Order certification presentation until it rivaled the finest doctorial thesis of any Ivy League school. Thirty-nine pages of magnificent prose covered every facet and every in-flight contingency of this SIOP sortie! This would be the certification that would not soon be forgotten; one to remember. It would be a certification by which all others would be measured in the future. This would so impress the Wing Commander and staff that when it was over they would stand and applaud.

Finally, the Chief of Current Operations told me he was tired of seeing me and it was time to plan my certification! It was he who had set me on this course of enlightenment by telling me I had to "memorize everything" and "don't read your presentation". I rehearsed my presentation for hours in front of a mirror. Every word of every page of the presentation was committed to memory. No, there would be no reading here.

By now you are asking yourself how this neophyte of international defense policy, how this mere second lieutenant could have transformed himself into the source authority of nuclear weapons employment. Easy, I memorized it! No small task at the height of ice fishing season in the middle of ice fishing heaven. Grand Forks is on the North Dakota/Minnesota border and Minnesota has great ice fishing. The Retired on Active Duty (ROAD) crew that constituted The vault staff lived to ice fish, so the vault closed each day by 1430. So, to have the necessary study time, I simply packed up whatever red covered manual and/or document I was studying at closing time and took it back to the BOQ room; no problem! My routine was to study in the vault until 1430; go to the base gym to lift weights and play racquet ball until 1700; head home to drop my stuff off, shower, and change for dinner; dinner and drinks at the O'Club until 1900; then back to the room to burn the midnight oil. That way I had the time to work on my presentation, read, and study all those darn red books.

The mission folders were all oversized accordion folded maps or documents printed on legal size paper – neither fit very easily into my briefcase so I rarely took any of them home to study, unless of course it was really necessary. I ate breakfast (for 25 cents) at the alert facility every morning at 0700 and then went to the vault to start studying. The vault staff was usually just leaving for the morning alert briefing so I would sign in whatever document I had signed out the day before and pull the next set of books. The vault staff was very meticulous about signing in and out documents when I took them to study, and I didn't want to do anything to upset those guys. They were the only people helping me. I was master of my world and, of course, doing a great job. Making new friends, learning new things, mastering new skills; life was good - until the Friday night before the week I was scheduled to certify.

That was the night the vault staff was inventorying the library and realized some of the red-covered books were missing. Apparently they had spent several hours looking for the misplaced documents. By the time the EWO Study Officer tracked me down at Happy Hour, he was quite agitated. Pulling me aside he confronted me. "Well of course I have the documents," I innocently replied, failing to mention that they were in my briefcase under the bed in my Q room. I had them signed out correctly, after all. In my ignorant bliss, I immediately retrieved them and drove them out to the alert facility, still unaware of the turmoil I had caused. The documents were inventoried and returned; further inspection noted that I had indeed completed the Sign In/Out forms accurately and correctly – it was just that the in/out times were measured by days! I had taken literally the comments about not referring to or reading notes and in my zeal to learn and memorize all this material without the aid of references; I had gone a bit too far. The entire episode was resolved that night, and I was instructed that in the future ALL material had to be in the safe whenever I left the area.

Proper OPSEC/COMSEC procedures were not emphasized strongly during the EWO certification process – an omission that was immediately corrected. A security violation was a quick way to the unemployment line and nobody wanted that. The incident was resolved and never spoken of again. The lieutenant colonels adjourned to a van in the alert parking lot for an adult beverage and to discuss their renewed retirement options. They quickly returned to their afternoon ice fishing routine. Changes were, however, obvious even to a second lieutenant; someone was always left behind in the vault to check in/out material and OPSEC/COMSEC was included in all future EWO study and certifications. As for me, I spent the first two weekends after being declared Combat Ready page counting and inventorying every manual and document in the vault. Looking back, it was a small price to pay for my ignorance. But, as I said, I was the first second lieutenant in the Wing in years so I guess they underestimated a second lieutenant's capacity to be naive.

Having survived that minor diversion, I set about preparing for my certification that was the following week. On Tuesday, I did my dry run with the Current Ops Chief, EWO Study officer, and staff at the vault. As I recall the normal EWO certification was scheduled for two and half hours divided into two parts; a formal crew presentation followed by questions/answers. Since I was briefing by myself, I planned for the entire period, see previous comment about second lieutenants. Following my dry run presentation all the Current Ops

Chief had to say was that it was "thorough". The Q & A dry run was a snap. The feed back there was "don't be too quick to answer" and "look up non-bold print responses." That was a departure from his preparation instructions, but I assumed he meant "show 'em you know where it's written". I also found out that an Alert Crew would be supporting me for the question/answer portion but were busy and could not meet me until the day of the certification. Anyway, I was deemed ready; next, "The Big Show."

Wednesday morning broke bright and cold across the northern plains. The certification was to be conducted in the War Room in the Command Post far below the Wing Headquarters building at 9 AM. I arrived early to ensure I could, in fact, find the Command Post in the building basement. After going down several corridors, I ended up in a large latrine with doors at each end. Figuring the bathroom was probably someplace close to the Command Post, I went out the other door and sure enough there it was....... "319th BW Command Post, Authorized Personnel Only". I had arrived. The people in the Command Post knew I was coming, of course, and knew about the certification. They were nice and friendly showing me around to the War Room; mostly they wanted to see the real live second lieutenant. One of the officers asked to see my ID to check my date of birth. Wow, time flies; he allowed how he had a car that was older than me.

The War Room was a long, narrow room with a glass wall on one side separating the Command Post council area from the War Room and a large map of the world on the other. There was a podium with a small peanut light on it, a table immediately to the left of the podium with sortie material and documents, and they all faced a long table with high back chairs marked for the various staff disciplines: Commander, Vice-Commander, Sq Commander, Current Ops, Command & Control, Defense/EW, Bomb/Nav, etc. First came the guys from the vault with the support crew from the alert facility. We were introduced, but they were a little leery about providing support to a second lieutenant and the exposure it brought them in front of the wing staff. As all that was going on, my anxiety was steadily increasing. I sounded great in the shower in my Q room, brilliant even; but this, this was the War Room! Suddenly the lights dimmed to almost darkness and somebody yelled, "Ten-Hut". It was show time!

The crew stood at attention at the table, I braced at the podium, and the wing staff entered the room. Ominously I noted that both Alpha and Charlie were coming to this show. The glow of the Command Post council area gave a strange hue to the room and provided a backlight to the staff which made them just dark outlines to me. The purpose of the small peanut light on the podium was very clear now as it was the only bright light in the room. "Be seated gentleman, ... begin". The time was 0900 and my life was over! "TRAINING, TRAINING, TRAINING remember your training." I cleared my throat and started. "Good afternoon sir, I am Lieutenant Dan........" my mind went blank, not confused, not scared, blank! I had no idea who I was (I had already unconsciously established I didn't know what time it was), and I took my first glance at my notes. Time as I knew it stood still; everything moved in slow motion. As I stared at the page trying to make out my last name, I saw a large red form appear on the page. "Staley" I said much too loud. "I am certifying the copilot's portion of Sortie ......" again, blank. Glancing down at my notes for the second time in as many seconds I noticed two things: the first red blob was smearing across the page, and there were two new large red blobs merging with the first. I immediately realized what was happening....nose bleed! In my collegiate days I had been repeatedly kicked in the face by an opposing defensive tackle who had an incredible disliking for me. That resulted in my nose being crushed on one side so it bled whenever the nasal passage dried out; North Dakota is nothing if it's not dry in the winter. My mind raced. "DON'T STOP," and I continued, again too loud, "Zero Two".

As I looked back up to make eye contact with the Commander and his staff, the expression on their faces was one of utter shock. To my left the crew navigator, who would resign his commission in a month and go back to medical school, rose and came to my aid. The human body contains six quarts of blood; I think I had four quarts on that first page. "Get him to the bathroom," Alpha growled, "and stop that @#$% bleeding!" The nav and the pilot led me out of the room. I swear I heard laughing as the door closed. Sometime later, with half a roll of toilet paper stuck up my nose, the bleeding stopped. My uniform was a mess, partly from blood, but more from the water thrown on my face. The pilot dried the shirt under the hot air hand dryer in the bathroom while the nav worked on my nose. Finally, put back together, incredibly embarrassed, and just a little shaken, I was led back to the War Room.

As we entered, I noticed the mood in the room was significantly different, the lights were up and the staff, while not smiling, were certainly smirking. "Let's get this over before I have to explain how a junior officer died in my CP," Alpha said, turning to look directly at me ... " NO BLEEDING," he growled and bit his cigar in two. "Cut to the chase Co," Alpha bellowed, "your certifying sortie two; presentation is over; it's time for questions". Silence filled the room. The pilot sitting to my left moved uneasily in his chair, unsure of what was going to happen next. Charlie spoke first, "Lieutenant, where is the #$%^& alert facility?" Sensing a trick here and remembering, "Don't be too quick to answer," I paused. "End of the runway, Sir," was all I could come up with. "Good" was the reply. Alpha spoke next," When you get in the airplane what seat do you get in?" Again a trick? "The copilot seat, Sir, on the ... (I had a 50/50 chance) right." "Good, don't ever forget that."

With that, Alpha looked at the EWO study officer and said, "This certification is over; Charlie will sign the paper work; put 'em on alert." It was over. The entire exchange took less than three minutes. I was Combat Ready!

Post script:

Rarely did being a second lieutenant pay bigger dividends than when I arrived at the 319th. Whatever I did was simply chalked up to, "What did you think? He's a second lieutenant." This would change in the years to come as the crew force got younger and younger, but not for me; I was the first. I found out later that regardless of what happened, I was not going to be allowed to present my 39-page EWO dissertation. The EWO Study officer had already briefed the Wing Commander and Director of Operations as to my preparation and assured them of my readiness. I think he gave them a copy of my presentation, and the sheer number of pages scared them.

One thing was for sure, it was a certification that "would not be soon forgotten."

It was "one to remember."

The Night the Balloon Almost Went Up

Doug Cooper

It was sometime in 1974 or '75 and the Great Southeast Asia Land/Air War had been over for enough time for all of us to get back into the alert routine at all of the SAC bases. As our ops officer said, "The war is over here now." He was so right.

Anyway, I had escaped the crew force and was a certified command post controller for the 7th Bomb Wing at Carswell AFB, Texas. I was working the "mid shift" which began around 2000 and ended at 0600. Normal duties during the mid shift were to try and stay awake so we could acknowledge the infrequent tests of the Primary Alerting System, SAC's main communication means to the wings and to the crews on alert. Nighttime alert force exercises were more or less not done due to safety considerations.

SAC had a schedule of states of readiness beginning at "normal alert" and ending with the crews taxiing their birds and taking off while awaiting the National Command Authority authorization to go to war. Each of those "postures" had an exercise routine. The controllers at SAC had the means to set off base klaxons and seize command post radios to expedite the process with the local command post controllers backing them up in getting the word to the crews.

Many alert force exercises ended up as "Chinese Fire Drills," a concept I never understood; however, these exercises often resulted in calls for message repeats, help with maintenance problems and other difficulties. To fully complete a taxi exercise required a lot of support and caution by those involved. It is dangerous to move aircraft weighing nearly one-half million pounds which held nuclear weapons. There were stragglers and problems everywhere.

Anyway, I was on the mid shift one night and had done all of the housekeeping chores and other requirements of the checklist when the Primary Alerting System went off. "Probably another test," I thought. Then I noticed that the klaxon light was flashing. "Holy s***," hollered my NCO controller. Then came the message. It instructed the crews to start engines, taxi to the hold line, and await authority to launch. This was not an exercise. We opened the real checklist and began to make the notifications. No one notified asked for a repeat.

Crews did just as they were instructed. There were no maintenance problems. No fires. No missing crewmembers.

After what seemed to be an eternity, SAC broadcast the message returning all units to normal alert. We caught our breath. The old man asked for some coffee. Fortunately, we had a fresh pot going.

We were never told what precipitated the situation. We thought it might have been a screw-up at the North American Defense Command (NORAD) command post. I firmly believed, after that episode, that, if told to do so, our guys would do the mission. No questions asked.

Winter of 1965-66 - Glasgow AFB, MT

Jim Bradley

It was the winter of 1965-66 at Glasgow Air Force Base, Montana, up on the "High Line." The airbase was just a mere 40 miles south of the U. S. and Canadian border. There the possibility of serious winter weather was always lurking, and so it was one day in the winter of 1965-66.

I was scheduled for my regular alert tour as the Electronic Warfare Officer on my B-52D crew. We were scheduled to pull three days on alert. The schedule required that I be in the Alert Facility and in the Briefing Room at 0730 for the standard weather briefing, any special notices from the wing, etc.—you know all this was normal stuff.

When my wife and I left my base quarters we were greeted with a lazy snow of large flakes. As we proceeded the wind began to pick up and the snow became more of a swirling eddy in the streets on our route to the alert facility. As we neared the alert facility the wind began to blow considerably harder and the snow was becoming wind driven to the point of starting to drift little fingers in the streets.

I got out of my car and carried my baggage into the alert facility, walking against the wind driven snow. My wife headed home, arriving just before the real blizzard began. She had low visibility all the way back to the base quarters. I made it to the facility, placed my suitcase in the bedroom, and proceeded to the Briefing Room for the briefing to assume the alert sortie we were assigned.

We completed the briefing and went to the aircraft to complete the crew changeover. As we exited the facility, we noticed that the aircraft were faintly visible in the blowing snow. The wind chill was colder than a well digger in the Klondike would be. We conducted the changeover in record time and went back into the alert facility. The crew we just relieved went straight home, finding it a bit of a challenge.

Later, we heard that the Air Police had issued a statement stating: "Wherever you are, stay there!" There were some holiday parties going on at the time at the NCO and Officer Clubs. This meant that some were not going home that night - or for a couple of nights!

On alert we were directed to stay inside. Telling us to do so was not necessary! Outside the building, the wind was blowing snow at 35 miles per hour - it was snowing horizontally. You could not see the aircraft on the first parking stub, let alone any of the other eight on the alert tree. The storm continued for two and one-half days.

In the alert facility the fuel oil used for heating the building had gotten cold enough that it congealed and the furnace went out leaving the building without heat. The alert mess hall cooks could not get home and stayed at the facility. They slept on the divans in the TV lounge. We were relegated to eating breakfast foods for three meals per day for those two and one-half days. There was no way for the food service folks to get any other food items to us - nothing was moving on the base.

After the storm had passed the only vehicles moving on the base were the tugs used to move B-52Ds around on the ramp and the snowplows. The Civil Engineering Squadron borrowed the big tugs to go to the housing area and re-light furnaces and thaw out water lines. The snowplows were clearing the streets, the flight line, runway, and the taxiways including the alert parking tree. When the storm was over there was a great feeling of relief. No lives were lost and no serious damage had been done. One automobile at the NCO Club had been parked facing into the north wind. When it was dug out of the snowdrift the entire area under the hood was completely filled with snow. The snow had been blown through the radiator and up under the hood so hard that it was packed into the space.

Dim Star, Bright Star—Broken Hearted Mission

Ken Schmidt

I was assigned to the 23rd Bomb Squadron, 5th Bomb Wing, in the late summer of 1980. Soon after my checkout as a radar navigator in the B-52H was completed in November, a Buy None hit and I was placed on a crew as a substitute radar navigator to fly the Buy None mission. As it turned out, I was flying with an excellent crew and in a fantastic airplane and I scored the best bomb scores of the Buy None - a triple snyc with scores of 100 feet each and a single release with a score of 150 feet. Not too shabby for a guy just transferring from the B-52D.

1981 turned out to be an interesting year for me at Minot AFB, North Dakota. It started out with being assigned to a new crew that was being deployed to Andersen AFB, Guam, for Busy Island. The deployment actually began on December 23, 1980. Busy Island was a Higher Headquarters special tasking for a show of force in the Indian Ocean as a direct result of the Iranian Hostage Situation of 1980. The tasking mandated that four B-52Hs be continuously deployed to Andersen for ongoing missions and Busy Island was supported by all the B-52H units on a rotating basis.

Upon my return from Guam, I was informed that I would be on one of only four B-52H crews throughout SAC selected to participate in five months of training which culminated in a six-week TDY to RAF Marham in England to participate in the last RAF Bombing Competition, Giant Strike 1981. This was to be followed by a TDY to Castle AFB, California, for Central Flight Instructor Course (CFIC) to become an instructor radar navigator. Finally, I was a participant in Exercise Dim Star/Bright Star 1981. Operation Bright Star was designed to be a series of biennial combined and joint training exercises led by American and Egyptian forces in Egypt. These exercises began in 1980 and were designed to strengthen ties between the Egyptian and American militaries and demonstrate and enhance the ability of the Americans to reinforce their allies in the Middle East in the event of war.

A little side note here: during the same 10-month time frame, I went through three crew changes. When I got back from the Guam deployment I was notified that a special crew was being "built" for the RAF Bombing Competition and I would be the radar navigator for that crew. I must admit that special crew was one of the best crews I had the pleasure to serve with. Unfortunately, when my crew returned from the competition, we were disbanded and I was sent to CFIC. Once back from CFIC and after local upgrade training was completed, I was placed on a new crew, albeit one without a pilot (but there was one currently upgrading and he would be the new pilot). During that timeframe, the crew was tasked to be part of the Bright Star "group" and for training, we had an outstanding IP from Training Flight (Maj Fred Bieber) as our pilot.

SAC had broadened the missions of the 5th and 319th Bomb Wings by developing the Strategic Projection Force. Following the 1980 Iran hostage crisis, SAC tasked the 57th Air Division, which had been on Minot AFB since the mid 70's, to organize the Strategic Projection Force. The 57 AD became the first division in SAC to implement the concept of using conventional strategic forces (bomber, tanker, and reconnaissance) in crisis situations worldwide. The 5 BW's B-52H fleet became the spearhead of the force, able to provide conventional warfare anywhere in the world.

The tasking for Bright Star '81 called for eight B-52Hs (four from the 5th Bomb Wing at Minot and four from the 319th Bomb Wing out of Grand Forks) to take off from their respective bases as four ship cells, to join up in an eight-ship cell at a pre-designated point and fly non-stop using in-flight refueling to a target zone in Egypt and return to the US. The training for this mission dictated that the crews participate in a practice training session that was designated Operation Dim Star. I believe that for Dim Star, there were a total of 12 B-52s (six from each base)...the four designated "players" from each wing and two spares. The missions for both Dim Star and Bright Star were preplanned for each crew and each crew had its own mission folder complete with high and low altitude mission maps and flight plans already completed.

Each aircraft was loaded with 27 live 500-pound retarded bombs for dropping at low level. Since cell procedures were standardized within SAC, the participating crews just needed to be in the "right place at the right time" for the cell join-up of the aircraft from the two bomb wings. The six B-52Hs from each base took off and did an enroute rendezvous to form a 12-ship cell and we flew to the White Sands missile range at which point each crew maintained aircraft spacing and timing as we entered a low level navigation route to the target area. Standard SAC procedures were used as we departed the low level entry point, proceeded to the low level navigation area, completed a nighttime TA calibration leg and then followed the planned course to the target area, dropped our weapons, exited the area and rejoined the cell for the flight back to home base. Contrary to what I had been told, I did not feel any change in altitude as the weapons were released - this had been my first time to drop a full load of 27 500-pounders—and it was a great experience. The Dim Star exercise worked flawlessly.

My crew was originally tasked to be one of the four main crews from Minot. However, a decision was made that since we did not have a "designated assigned pilot" as part of our crew, we got bumped from the lineup and would be relegated to a spare status for the Bright Star mission. Needless to say, my crew was disappointed since we had trained hard and had the successful Dim Star mission behind us. Anyway, the day for the mission arrived and all the tasked crews met at the squadron, got the mission and weather briefings, collected all the associated mission gear to include flight meals, extra water and coffee, Tac-comm gear, etc., and proceeded to each individual aircraft for preflight. Everything went according to plan to this point: aircraft were preflighted, weapons were preflighted and pins removed and stowed, and we got ready for engine start. Engines were started, equipment turned on and we waited to taxi.

At that point, one of the main aircraft had a problem with the pilot's communication system and it looked like that they would have to drop out. It was too late to do a bag-drag, so the decision was made to let us (the spare crew) launch and go with the cell. Up to that point, my crew's thought process had been relegated to shutting down engines, calling for transportation to pick us up, then to do normal post-flight procedures. Then our mindset changed and we were really looking forward to launching and completing the mission. To a man, we said "Hell, yeah, let's go." However, just as we were getting ready to taxi, the other crew reported that the communications problem was "fixed" and they would launch. Our hopes literally fell to the taxiway and we waited, with engines running until the "favored four" took off. Dejected, we completed post flight checklists, shut down engines, got the ground transportation, returned all the gear, and went back to the squadron.

The Bright Star '81 mission was a success; the four aircraft from Minot joined up with the four aircraft from Grand Forks, flew their mission, successfully dropped their weapons on time/on target and returned to their bases after a 31-hour flight. I remember seeing national news footage on TV showing the B-52s flying low over the target area and seeing the bombs deploy and thinking "that could have been me and my crew".

There is a postscript to this little ditty. A few days later, I was talking to the radar navigator on the "affected" plane with the communications problem. It turned out to be a bad connection in the pilot's helmet, rendering his intercom system useless. The pilot made the decision to fly the entire mission using his headset.

No mention of the exact nature of the "problem" or its "solution" was ever mentioned on the ground over the radios and therefore there was no input from the Wing Commander, the Vice Wing Commander, or the DO in the decision making process.

The crew that made the decision to fly was a Stan/Eval crew.

Talk about double standards.

Plotting the DEW Line

Jim Bradley

There are many stories of crewdogs doing famous or infamous things, without thinking, that gets them into trouble. This is one of those stories.

When the Strategic Air Command decided that the Soviet submarine launched missiles made some US bases vulnerable, they decided to disperse the bomber fleet. So in February of 1961 the 326th Bombardment Squadron, equipped with B-52C/D aircraft, was moved from the 92nd Bombardment Wing at Fairchild AFB, Washington, to Glasgow AFB, Montana, as the nucleus for the organization of the 4141st Strategic Wing.

In its first year, it became the top wing in Fifteenth Air Force. The 4141st Strategic Wing at Glasgow AFB was inactivated on 1 February 1963, and its aircraft were transferred to the 322nd Bombardment Squadron that was assigned to the 91st Bombardment Wing at Glasgow.

I was stationed at Glasgow and assigned to the 91st Bombardment Wing (Heavy), 322nd Bombardment Squadron, as an Electronic Warfare Officer on a B-52D. Although this story happened before my assignment to the unit, it is about one of my crewmates on Crew E-31 to which I was assigned.

A study in 1952 had also identified the vulnerability of the United States and Canada to air attack from bombers of the Soviet Union. To counter this threat a series of radar stations were constructed in the northern Arctic region of Canada and along the north coast of the Aluetian Islands, Faroe Islands, Greenland, and Iceland. The combined stations were known as the Distant Early Warning (DEW) Line, and during the Cold War served as an early warning detection system to monitor Soviet aircraft flights.

As earlier stated, the sub-launched Soviet missiles also reduced the time it would take to respond, so the Strategic Air Command devised the plan to improve the posture of national defense by flying regular airborne alert missions over the Arctic. Having aircraft already airborne would greatly reduce the time required for a retaliatory strike should the Soviets decide to attack. These flights were performed by nuclear-armed B-52s and were called "Chrome Dome" missions.

One of the Navigators in the 322nd Bomb Squadron became famous for plotting the location of these Distant Early Warning Line (DEW Line) radar sites over the course of several of the airborne alert flights he flew.

He had plotted them on his flight chart and eventually proudly showed some of the charts to several of the squadron's other navigators, spreading around the knowledge he had collected on the sites. Later, after flying one of these airborne alert missions he had to turn in one of his charts and was subsequently called in and "counseled" as to what the plots on his chart represented. He explained that they were the bright returns on his radar scope which he saw as the flights transited the polar ice off the northern coast of Canada.

He was informed that because he had noted all these site locations, his charts were now "Classified" and that all of them would be confiscated.

The exact locations of the DEW stations were a matter of national defense and not something that could be treated lightly.

Cold War Duty – '66 to '68 in a B-52H

Walter Want

I was a 1964 graduate of the ROTC program at NMSU and was headed to Willie (Williams AFB) to earn my coveted pilot wings. I finally got it done after a fall off a bar stool in Nogales set me back six weeks. I rode in the back seat of the F-5 with orders not to allow the foreign pilots to hit the ground or tour the Grand Canyon or the main streets in Vegas at night. I was soon headed to fighter pilot school at Nellis AFB, Nevada, - or so I thought. A big need for pilots by SAC changed my dreams and aspirations. I was sent to Castle AFB in Merced, California, then Carswell AFB, Texas, and on through survival training at Fairchild AFB in Spokane. My final destination was B-52Hs in Grand Forks AFB, North Dakota.

Wow! - a new wife, new airplanes, and winter weather. We were greeted just south of GFAFB where we were stopped by the highway patrol and ushered into a large culvert to await an oncoming tornado. Our yellow VW bug was left topside with everything we owned inside.

In August of '66 I checked in and there's no base housing available for new guys at the 319th BW(H) and the 46th BS. Came home with a full load of cold weather gear and we had a frolic, all dressed up in our new furs and insulated undies. A long look for a honeymoon apartment put us in an 8' x 50' mobile home near the railroad tracks on the way to Winnipeg just before the first storm of the season. That was followed by my airplane check out, targeting, and all that other stuff designed to fill a new CP's head prior to winter alert duty as a very Cold War Warrior.

Things went well though and we adjusted; easily done when you are newlywed and the FNG on base. I was assigned to a crew, and made rapid upgrades to ICP and on a select crew in short order. Then to my surprise, my upgrade to pilot training started within a year. I won an award for a project called Busy Figure II, Boeing service and performance testing. We flew a complex profile, photo documented, in numerous airplanes and the rumor was that the H-model fuel and performance curves were never inflight verified. Numerous changes came down to our Dash Ones after that.

I have early memories of 36-hour missions over the North Pole, the Northern lights, air refueling contrails from Alaska for a thousand miles, tired butt, bad food, and an alert message for the primary crew to occupy their assigned seats for a pending message got our attention.

We were selected as the demo crew for authentication of a message to demonstrate our combat readiness. You got it! It was hours of boredom punctuated by moments of terror. Were we really going? Or was this just another test? The warble came in and we did our authentication thing and impressed some politician that crew S-03 was alert over the North Pole.

There were long cold rides and beautiful scenery and many times you could look back at the Southern Lights. The best thing about it was you got home time and lots of hours those months.

The bad thing was you got to look forward to doing it all again.

Friday Afternoon Follies at Carswell

Ken Schmidt

I don't remember the exact date but for all intents and purposes, it could have been Friday the 13th. It was definitely a Friday in early 1980. The flying day started out fairly typical: gather at the 9th Bomb Squadron for an "0 dark 30" morning flight, pick up personal equipment, board the bus and go to base ops to pick up the TACCOM materials, get the weather brief, gather flight lunches, water, and coffee, and get to the B-52D aircraft. Once at the aircraft, the pilot and crew reviewed the maintenance documents and quickly went through a mission recap then we offloaded our equipment from the bus to the airplane, completed our walk around inspections, completed preflight and then started engines and turned on equipment.

We actually had an on-time takeoff and had a near perfect mission that consisted of air refueling, a low level navigation leg and a couple of bomb runs, climb to altitude, complete a celestial navigation leg and return to base for an hour's worth of touch and goes for the pilot team. All was great until we got back into the pattern when we heard "Mayday, Mayday, Mayday" over the radios. Two T-33s were on takeoff roll out of Carswell when one of the T-Birds somehow managed to jettison his wingtip fuel tanks which in turn caught fire when they made contact with the ground. Needless to say, the runway was closed and all aircraft in the pattern were directed to orbit at assigned altitudes.

This being Carswell, (two bomb squadrons, a tanker squadron, CCTS and CFIC) there were several aircraft in the pattern. After burning holes in the sky for over two hours, fuel was getting to be a concern for several of the aircraft with the high probability of diverting to Dyess. However, there was one enterprising KC-135 crew that was airborne who made the call "if any of you guys need gas, give me a call and we can offload some to you". That was not needed as at that point in time and all airborne crews were given the go ahead to land. All the fires were out, the FOD cleared, and all was right with the world and we got to land.

However, as Paul Harvey would say, "Now the rest of the story." After landing and taxiing back to parking, we did the normal post-flight activities: shut down the equipment and engines; offloaded the plane;

boarded the bus and finished the maintenance write-ups and gave the 781 to the crew chief; returned to Base Ops to turn in the TACCOM materials and water/coffee jugs; then returned to the squadron.

Just as we were getting ready to debrief and grab a beer, the Squadron Ops Officer stepped in and said "Don't drink anything, you're going on alert". Needless to say, our collective jaws all dropped and we asked "What the hell is going on?"

He said, "You will find out when you get on alert. Now get going; collect all your alert gear, and report to the alert shack by 1700". After gathering out alert gear, we drove our cars out to the alert crew parking lot where we were met by another crew's gunner, loaded our gear into the truck and proceeded to the alert pad. Questions were asked but no answers were given.

When we got to the pad, we were finally briefed as to what was going on. It seems that just prior to the time the T-33 dropped the tanks, the Wing got a no-notice klaxon. All the alert crews responded to their aircraft and did their alert "thing": started engines, copied and decoded messages, and responded to the messages. All of the crews did them correctly - all but one.

That crew had incorrectly decoded the message and started to taxi, then the crew rechecked the message and discovered that it was actually a NO TAXI exercise. When they stopped the aircraft, they had just barely taxied out of their spot, but they did TAXI on a NO TAXI exercise. That was enough for them to get pulled off of alert and get themselves decertified for EWO operations.

We were the fortunate "Only Crew Available" to replace them. We got to spend Friday night on alert, but were relieved the next morning by the old crew who had managed to get themselves "mission ready" again after being recertified by the "Wing King."

Never a dull moment at Carswell.

RAF Blue Steel, A Reflection

Mike Radke

After the closure of Glasgow in 1968, I was assigned to KI Sawyer to fly H-models. With so many base closures taking place and personnel being reassigned there was an influx of new personnel to form the new crews. I do not recall much about the first crew I was on, but regardless, we fell into the usual crew routine of three sorties and ten days of alert per month, plus ORIs and Chrome Dome flights.

Sometime after the crews were formed we began to rotate into the Arc Light circle. About a year and a half later my crew's rotation time came around. I was the only member of that crew who had previous Arc Light experience. Charlie, a captain, was the pilot, and Barry, also a captain, was the copilot. Mike, a major was the RN; Walt, a captain, was the nav; and George, a first lieutenant, was my EW. He had the moniker of "George of the Jungle" for some reason. I really enjoyed that crew and we had a lot of professional rapport. I do not recall any major events that happened during our tour – just the normal Arc Light happenings. When we completed our tour we boarded a tanker headed for home and had a layover in Hawaii. The copilot's family met us there and took us all to a luau. It turned out his family owned a fleet of fishing boats and many pineapple plantations and we each went home with a case of pineapples.

When we got back home to Glasgow we fell back into the typical crew rotation of missions, training, and alerts. We were home only a few months when we were selected as the designated crew to go to England for the RAF's Blue Steel Competition. My pilot declined the offer because he was getting married. An agreement was reached that the wives could accompany us there, and they could reside at the Base Commander's home which he wasn't occupying. I can't recall why we had an EW change, but it took place after an ORI, so Will became our EW for the competition.

The officers resided in the BOQ and I stayed in the NCO quarters. It was a totally new experience since the Brits are really different in their military mannerisms and messing was a total experience. Our sponsor was the 214 Squadron, an excellent host.

We started our training and flew almost every day. Crew responsibilities were somewhat different. We had a spare navigator and the EW assisted the nav team more. My responsibility was being the radio operator and there was a constant stream of communication. A lot of our off-duty time was spent touring castles or checking out antique shops. It was quite a chore navigating the narrow English roads in a Dodge Power Wagon with the steering wheel on the wrong side. Then we had fuel problems, and the crew chiefs were always siphoning fuel out of equipment. There was a lot of shuttling back and forth to RAF Lakenheath for fuel and ice.

When Easter came the tanker crews were going to Spain. Since the Cold War was still on we were still designated nuclear capable we couldn't go. The wives were busy touring and doing their own thing, so we decided to take in London and headed off to the train station and on to London. I housed with Mike the RN there and we had different interests than the rest of the crew. The night spots we found in London were either in the basements or upstairs and the only access was by elevator. We joined the rest of the crew for a drink once and then left to tour on our own. Apparently the rest of the crew didn't understand the relationship with the bar maids, and after a few hours of drinking they found that their bar tabs were 100 pounds each. Entertaining the bar maids was expensive.

Back at Marham the night life was centered in the NCO mess. The mess was open to the public so the evenings were quite crowded, especially after it was discovered that we were hauling in ice so we didn't have to drink warm ale. We sort of lived on fish and chips until Easter. We also dyed a lot of Easter eggs and hid them for the kids in housing, which was fun. When Easter was over, it was back to reality and a little more training. The competition began with a day sortie, followed by a night sortie, and finally another day sortie. When the competition was over we found that we had outscored everyone else. It was quite an affair to win the Blue Steel, but fame is always short-lived on a SAC crew.

Upon our return to KI Sawyer, we got Charlie married and it was back to the normal routine. On our second alert tour after returning, Charlie got really sick halfway through the tour and we had day to day substitutes. Charlie was eventually grounded and was assigned to the Civil Engineers. T'was a real shame. Barry, my copilot of Japanese descent, always wore his hair long and continued to get write-ups at the PX but the commander ignored them.

Life of the Crewdogs went on. C'est La Vie.

Getting the Gas – B-52D Aerial Refueling

Tom Jones

Aerial refueling was the toughest skill I had to master as the pilot of a B-52D Stratofortress. Many of you were instructor pilots in the B-52 and taught aerial refueling to many a novice like me. You dealt with new copilots getting a little introduction to BUFF formation flying and refueling, experienced copilots getting some time on the boom in preparation for upgrade to aircraft commander, and qualified AC's expanding their performance envelope behind the tanker. Hats off to you gentlemen, who always flew effortlessly, calmly communicated your experience and skills, and exercised limitless patience while teaching those of us new to the B-52. That you could successfully create a pilot with the skills to stay with a KC-135, flying at 255 knots, at night, in turbulence, during the aerial transfer of anywhere from 10,000 to 100,000 pounds of JP-4, is a tribute to your patience and teaching ability.

To achieve and maintain a stable position inside the tanker's refueling box, or envelope, just below and aft of the tanker's tail surfaces, pilots had several visual and even audio references. Lateral clues came from lining up directly behind the tanker's fuselage. As the range closed to a couple of hundred yards, you could line up with the big yellow stripe painted down the middle of the tanker's belly.

I'd get the right vertical alignment by flying so as to put the boom nozzle in line with the tanker's nose. By flying slowly forward and up while maintaining that line, I'd arrive finally at the pre-contact position, 50 feet astern of where the boomer could plug me.

Forward and aft spacing was the most challenging. One determined pre-contact position by lining up the -135's engine pods at known locations on the BUFF's left and center windscreens. Triangulating on the tanker's engines gave you the clue you'd arrived.

Once cleared for contact, I squeezed in a bit of throttle and crept forward to stabilize just aft of the tail surfaces, but there was no easy cue for exact positioning. Often the boom operator would flash his director lights, on a panel up under the nose. A flash of the "F" light meant come forward a bit; a flashing "D" told you to drop down a couple of feet. If those cues weren't doing the job, a boomer might key the mike and say "Up five," to coax you into the center of the envelope. Tiny corrections were crucial here, but it took me several refueling sessions to learn to moderate my sometimes clumsy inputs. I extend my apologies to the many boomers out there whose nerves are still jangling some 30 years later. The rare "Breakaway!" calls you gave me were well-deserved.

What saved me were my good instructors, particularly my AC, the late Capt Benny Jubela of the 20th BMS. Although I was eager to learn how to refuel, and eventually did it well, handling the B-52 behind the tanker was one of the most difficult flying tasks I ever faced. The crucial task of drawing 50 or more tons of jet fuel through the tanker boom, absolutely required in combat to reach the target and return to home base, was a daunting one.

First, the B-52 itself weighed about 210 tons at the start of a typical training sortie rendezvous and refueling; this usually occurred about ninety minutes after takeoff. That's a lot of inertia, and the aircraft does not respond instantly to throttle or control yoke inputs. Anticipation and small corrections were essential to stabilizing the bomber in the contact position behind the tanker.

Second, in the early 1980s, the BUFF's refueling autopilot mode was not always reliable. This mode could be engaged in the pre-contact position; it provided a "power steering" mode for the pilot, flying through the control yoke. The autopilot reduced the yoke forces and automatically trimmed the aircraft to maintain the commanded attitude even as fuel was pumped aboard from the tanker, shifting the aircraft center of gravity. When working right, it reduced the pilot's workload and was much less fatiguing.

Unfortunately, the AR mode was sensitive to turbulence and often disengaged while trying to maintain position behind the tanker. When the autopilot kicked off, it gave you an automatic disconnect from the boom and dumped you into manual flying mode – great news when you were bouncing around in the weather and fighting to stay within the refueling envelope a few feet below the KC-135.

An excruciating moment for me, early in each refueling, was the initial creep up from the pre-contact position, 50 feet back, to the contact position in the refueling envelope. It still makes me cringe to remember sliding the nose of that big bomber up under the tail surfaces of the -135, and watching the end of the boom make a beeline toward our center windscreen. At just the last moment the boomer would deftly pop the nozzle up and over our windows and toward the refueling receptacle just behind the pilots' seats. The 250-knot slipstream past the nozzle roared in our ears as the boom scraped above our heads by a few inches. Each second turned to hours as I tried to hold a stable position beneath the boom operator's window. I had few visual references for precise fore and aft position until the boom slid home – "Plug me now – please!" At last, I heard the loud "thunk" of nozzle seating in the receptacle – and the tanker's director lights flashed on to give me accurate up-down, fore-and-aft information. Relieved, I finally knew where I was in the envelope and could make accurate corrections.

A third challenge during refueling was that without the autopilot, the B-52 could quickly outwrestle and exhaust a new or tense pilot. Because of the big bomber's inertia, the pilot had to anticipate deviations or trends out of the envelope limits, and make corrective control inputs before the airplane built up too much momentum to be brought safely back in line. Without the autopilot, while trying to stay centered in the envelope, it was easy to put too much yoke input in and overcorrect—you would then have to counteract your first input to avoid overshooting. Pretty soon you were horsing the yoke around, fighting yourself more than the external forces acting on the airplane. Many times early in my AC training, I found myself constantly twisting the yoke or jerking it up and down. That not only bounced the airplane around (and frightened the boom operator!), but quickly took the starch out of my left arm. Once you started fighting the airplane like that, you could only delay the inevitable disconnect or breakaway called by the boomer. It took a conscious mental effort to make small inputs, then wait and see if the correction was sufficient before putting in the next yoke movement in pitch or roll.

At the Carswell base gym, I started lifting weights, trying to build up some left arm muscles! And I worked harder to control the sometimes panicky urge I would have on the boom to throw in six inches of yoke movement, trying to catch a minor burble or deviation in position. Once I learned to take a deep breath, relax (a bit), and ride out most of the small excursions, I found the airplane's inertia could help me: it resisted rapid accelerations, and would give me time to assess the bumps and my own mistakes before making a correction.

During my aircraft commander check ride, I was doing my best to stay plugged in behind the -135 as we deviated east and west across Louisiana, dodging summer thunderstorms and all their attendant turbulence. I'd had a few disconnects already, and was really sweating to stay in position despite all the bouncing around. The autopilot was AWOL. Seeing my struggles and the limited amount of time we had on the track to get my 30 minutes of boom time, my instructor/check pilot took pity on me by providing a hint I would use on other challenging flights. He had me turn over throttle control to the copilot. With the co-gently managing our fore and aft position, and me handling only the pitch and roll, I could settle down on the yoke and make smaller, more controlled inputs as we and the tanker shuddered through the clouds. Like magic, I now had a right hand to help manage the yoke instead of my burned-out left arm; the workload was then something I could handle. During exercises like Global Shield, with fuel transfers of 100,000 pounds or more, that technique was a real ace in the hole.

I've heard from B-52 pilots who've flown the other models that the D-model provided the pilot a significant advantage in aerial refueling. The D, like the earlier models, had control surfaces driven not by hydraulic boost, but by air loads generated by cable-controlled trim tabs. For roll control, the D had small but effective ailerons halfway out on each wing. These would handle fine roll movements; when greater roll authority was needed, hydraulically driven spoilers atop each wing would deploy to create the necessary lift differential across the span. During refueling, a D-model pilot could make small roll corrections using aileron alone, commanded by the first few degrees of yoke travel. Often this was enough correction to stop the unwanted motion. But if you over-controlled and threw in too much yoke, a spoiler would pop up on the downward-moving wing, dumping lift abruptly, perhaps more than needed.

The raised spoiler would also cause just enough drag to yaw the bomber in the AR envelope, something else to correct for. B-52G's and H's had no ailerons, just spoilers, so D drivers and their ailerons had a bit more precision to call on behind the tanker - if they were skilled enough to take advantage of it.

When our time on track was up, we'd taken on the needed gas, and the copilot had had a chance to fly formation in pre-contact, it was time for blessed relief. I'd make one final contact, stabilize for a minute or so, say "so long" and thanks to the boomer and tanker crew, then squeeze the yoke-mounted button for a disconnect. Dropping off the boom, I'd slide the eight throttles gently aft and nose down and away from our flying gas station. Back at our cruise altitude and speed, on autopilot and on heading for low level, I could feel my sweat-soaked flight suit start to chill. The nav team had sent up a cold Dixie cup of lemonade for me – no beer was ever wetter, colder, or more satisfying than that slug of lemonade as the tanker crew departed, their mission accomplished.

By the end of my BUFF career, I'd had enough good ACs and instructors to become a decent refueler: proficient enough that I could match contacts with my IPs, staying with the tanker on seven engines, through non-autopilot turns, sliding easily back and forth in the envelope while easing the boomer's worries: "That guy hanging on the other end of my boom just might know what he'd doing."

On one of my last flights in the D-model, enjoying the challenge of staying behind a particularly sharp -135 crew, I remember saying to my IP: "I can't believe they actually pay us to do this." He smiled back at me; he knew exactly the satisfaction that I was feeling. It's one of those Air Force benefits that I still savor.

The B-52 and the Space Shuttle

Tom Jones

The Boeing B-52 and the Rockwell/Boeing Space Shuttle are not usually spoken of in the same conversation. But in my life they were inextricably linked, with hard-won experiences in the BUFF serving as invaluable preparation for my training and flights in the space shuttle program.

To say that I was disappointed at being assigned out of Air Force pilot training to the B-52D in 1978 would be an understatement of the first order. At the Air Force Academy, my sights were set on a career with NASA and the space program, and Air Force flying and test-piloting were my vehicles to achieve that. I would first become a fighter pilot, then take that high performance flying experience into test pilot school, and on to the space program.

Reflecting this plan, my "dream sheet" out of pilot training put an F-15 Eagle as my top choice, with other fighters filling out the list. However, the omnipotent Air Force personnel system computer had other plans. I wound up as a copilot in a B-52D out of Carswell.

One of my IPs at Vance gave me some mature advice to soften the dejection I felt. Even SAC deserved good pilots, he said, and as a former BUFF pilot himself, he thought I would enjoy the added conventional bombing mission assigned to the D-models. He also emphasized how satisfying it was to become proficient in handling the B-52 in its varied flight regimes: takeoff and landing, air refueling, low level bombing, high-altitude bombing, and air-to-air maneuvering.

That helped – a little. When I arrived at Carswell I was immediately impressed: a D-model launching out of Carswell at six a.m. thundered over my hotel and nearly shook me out of my bed. The shriek of eight water-injected J57-P-19W engines convinced me the bomber was about to rip through the lobby itself.

Six months later, after navigating the classrooms and simulators of Combat Crew Training School, I was a new copilot on alert with the 20th Bomb Squadron. The true essence of the B-52 crewdog job became apparent: teamwork. If I did my job and helped the crew do theirs, together we would succeed on each mission. It was impossible to look good as a copilot without the help of my crew, just as they seemed to need the green rookie in the copilot's seat to meet their own – and the crew's – training and deterrence objectives. Never did that principle come into sharper focus than when checking out my first alert bird with my new crew. Crawling with the nav up in the bomb bay around a covey of B-28 gravity weapons, I got it – this nuclear deterrence business was extremely serious, and the only route to success and survival was to work together. Only together would we have a chance at meeting SAC's standards and be ready to put bombs on the target.

I respected my first aircraft commanders: J.J. Wright, Lee Gettemy, and Benny Jubela, not only for their flying skill, but for their ability to pull their crews together. Sure, we were all professionals, but we had individual weaknesses as well as strengths. The pilots I knew were able to use a mixture of personal example, good humor, and almost subliminal appeals to professional pride to get the best performance out of our team. In this approach to leadership, I saw that my own good performance meant little unless all of us supported each other and helped the crew succeed. And when we did succeed, it felt very good. It's impossible to underrate the satisfaction of being a respected and valued member of a small group. The esteem of the crew was what I valued, far outweighing any incentives offered up by SAC, the squadron commander, or the Wing King. More than advancement or salary or individual reputation, I wanted to be known as a copilot, and then an AC, who was part of a damn good crew.

Teamwork was the most satisfying take-away from my five years in SAC and the B-52. The teamwork moments I remember are many: practicing stopwatch bomb release timing for the RN, hacked from a low level visual landmark; splitting up the air refueling flying task, particularly in turbulence, with my copilot; plunging into an emergency descent when my gunner lost tail pressurization; getting a dead-on steer to the tanker in heavy weather from my trusty nav-team; as a young copilot, savoring the bomb runs, approaches, landings, and refueling time handed to me by an AC willing to teach his inexperienced copilot; sharing jokes from tail to cockpit and back again as we cruised the stratosphere on cel nav legs; putting the EW and nav team in the front seat for some flying time during those long high-altitude runs to Labrador and Greenland; jinking to put our F-101 pursuer into the gunner's sights on intercepts over the Gulf of Mexico; laughing on a night low-level in Texas as the gunner lit up road traffic with his Aldis lamp; calling up the downstairs crew to come topside and marvel at St. Elmo's Fire and the brilliant Milky Way on those long night flights; and getting a shack on an ORI to put the crew high up there on the scoreboard. Those were some satisfying moments, and without the crew coordination and professional teamwork we practiced, we could not have been confident in our ability to fly and survive in a combat environment.

Seven years after leaving the Air Force and the B-52, I started crew training for the space shuttle at the Johnson Space Center in Houston. I was flying again, to be sure, in the two-seat T-38 trainers, but we were aiming for higher flight levels: low Earth orbit, at least 110 miles up. What I learned very quickly in the space shuttle mission simulators was that our instructor teams deliberately searched for the seams in our crew coordination and teamwork.

In what we called "generic" sims, with a pick-up team of non-assigned astronauts, one veteran flier might put on a strong performance of flying and emergency management, but it was simply impossible for a shuttle crew to get the job done flying solo. The instructor squad would overload one of the crew with successive failures, immobilizing him. Next thing you knew, say on reentry at Mach 15, our task-saturated crew would miss some subtle failure cue, the orbiter's guidance, navigation, and control system would suffer a debilitating malfunction, and with a shudder and jolt, the orbiter simulator jerk to an abrupt and dismaying halt. We were dead.

During the debriefing, we would discover our shortcomings and realize that cockpit resource management was the key to survival. During shuttle launch, you had about 15 seconds to recover from a serious malfunction before the ship went out of control and broke up. On reentry, you might get 30-45 seconds to react. To stay alive, we had to talk to each other, to step in and help out our buddies when time was tight, emergency procedures overlapped, and immediate action was critical.

After five years in the BUFF, I knew this. But what I hadn't known while flying for SAC was that teamwork was not a universal skill. Some astronauts came from academia or industry, others from the single-seat fighter world. Using the talents of all the crewmembers in the cockpit was not instinctive; we had to learn how to work together, and practice, practice, practice.

The teamwork paid off on STS-68, counting down to launch on Endeavour on August 18, 1994. The three main engines rumbled to life at T-minus-six seconds, surging up to full power just before zero, and lift-off. I was strapped into my center seat on the middeck, just to the right of classmate Jeff Wisoff, as we felt the orbiter rumble and shake under the thrust of a million pounds of liquid-fueled thrust. Out the hatch window I could see the gantry apparently sway — it was actually Endeavour "twanging" under the thrust. I mentally counted: 5... 4... 3... 2... 1... waiting for the giant kick from the boosters' ignition.

Instead, the Master Alarm blared in our headsets as the three main engines fell silent. Terry Wilcutt declared "Right engine down", and he and Mike Baker swung into their abort checklists. Jeff and I threw off harness straps and prepared to roll out of our seats, onto the middeck's back wall (the temporary "floor") and heave the hatch open for an emergency egress. We might even have to hit the slidewire baskets for an escape to the blast bunker a quarter mile — and a long zip down the slide wire — away.

Instead of liftoff, we were left swaying atop the orbiter as the launch team announced on the radio an RSLS (redundant set launch sequencer) abort – an automatic shutdown due to some as-yet unknown problem. Launch control soon verified we had no fire and no explosion risk. The engines had shut down at T-1 second, due to an overheating liquid oxygen turbopump on main engine #3. Its discharge temperature had violated redline limits; had we launched with that violation, we might have lost that engine right after liftoff, sending us into a very hairy Return to Launch Site abort.

Commander Mike Baker and pilot Terry Wilcutt wasted no time in completing the abort checklist, while flight engineers Dan Bursch and Steve Smith backed them up on the flight deck. Downstairs, Wisoff and I readied the hatch for opening, then settled in on intercom to wait for the ground crew to come out and open up from the White Room. We debated (in colorful terms) how long our mission delay would be — "We should have been gone!" Jeff lamented. When the ground crew arrived, I was sitting on my middeck seat back, legs dangling, munching a peanut butter and jelly sandwich the crew quarters staff had packed for orbit. We had been scrubbed, but thanks to the launch control team and our quick response, we were alive. We eventually launched six weeks later, on September 30.

That outcome wasn't a given. We had sharpened our cockpit emergency skills for months in training, and our ability to react quickly and properly had been honed to a sharp edge. That training began for me in the right seat of a B-52D, 16 years before.

What the Heck Is a MITO?

Dick Gaines

MITO is an acronym for a minimum interval takeoff. Each of several B-52's roll across the "hold line" just prior to taxiing on to the runway at an interval of 12 seconds. In wartime, this allows the SAC Alert Force to launch with an absolute minimum delay. It requires a high level of skill, especially in the early model B-52's that used water injection to boost power to the engines. Those models created a tremendous amount of black smoke which made for some anxious moments for the second, third, and fourth, etc, aircraft to takeoff. Visibility was cut to 200 feet or less.

In the event of an aborted takeoff, the aborting aircraft had the primary duty of initiating the abort and calling "ABORT, ABORT, ABORT" over the common radio frequency. The Air Base Control Tower would then pick up the abort procedure with a continuous call of (example) "THIS IS WALKER TOWER ON GUARD----ABORT, ABORT, ABORT".

Hopefully when the dust cleared all of the B-52's would be very close together, drag parachutes deployed, and rolling out to the end of the runway to figure out who had the abort problem. Once the drag chutes are deployed, flying is over for the day for those airplanes. In wartime they would launch without the drag chute.

Now for the true story (they all are).

It was November 10, 1964. My crew was leading a three B-52 MITO at Walker AFB, New Mexico (Roswell). I was copilot on the crew and Maj Angus Tidwell was the aircraft commander. The other two crews in the aircraft behind us were highly skilled Standardization crews, one with the SAC Wing Commander aboard.

Everything went very smoothly through preflight and engine start. We all taxied to the runway, completed our radio checks and at the exact planned time of takeoff we started to roll for takeoff, followed by the other two bombers, each 12 seconds apart. We lined up with the runway, initiated water injection and pushed the eight throttles to the stop.

Major Tidwell called "70 knots NOW!" for the radar navigator downstairs to begin timing to the "decision speed" which was around

118 knots. At about 110 knots I saw a bright red flash from the right wing and immediately saw the exhaust gas temp gauge for number seven engine start to come up.

Just as the RN started the announcement of "coming up on S1 ...," I made the decision to abort the takeoff. I had no time to tell the aircraft commander, so I pulled all the throttles to idle, and called "ABORT, ABORT, ABORT" on the interplane radio frequency. Walker Tower picked up the call and continued it until the other two B-52s behind us aborted. I remember clearly seeing the other two bombers right on our tail as we turned off at the end of runway 26.

At that point everything and everyone was safe except for my Air Force career. There was no fire on the right wing since we had shut down the number seven engine and pulled the firewall fuel shutoff valve. I was either going to get patted on the back or kicked a little lower. Thankfully it was the former. I sweated bullets until the Supervisor of Flying found pieces of our engine scattered for 4,000 feet down the runway.

Note: Bunny had a tiny piece of the engine placed in my medals display case which I have had on my family room wall in Maitland, Florida since 1984.

A Close Call for an Alert Crew

Walter Want

That first winter At Grand Forks AFB was amazing. I got up one morning and it seemed HOT. The temperature had risen from minus 32 to 16 degrees above zero and you could really feel it. The yellow VW Bug was all but frozen to the ground. I had to take the battery into the house, warm the oil with a heated dipstick, and have a hair dryer in the car to de-ice the windows. I got the hose out and attached it to the kitchen sink in our new mobile home and ran it out the window. I was going to wash the car during the heat wave. I sprayed it down good and returned inside to get soap in a bucket.

What a surprise I had when I returned a few moments later. The Bug, "Mellow Yellow," was absolutely frozen to the ground. It took a major effort to thaw it out so we could use it. The foam around the doors had already pulled loose and the windshield wipers were frozen and cracked. I never could keep a decent blade on that thing and ice just chewed them up. The Bug was not known for its cold weather heating capacity.

Following surviving a howling roof-rattling blizzard in our little mobile home one morning, the phone rang at 5 AM to announce that all alert crews were to report as scheduled. On time check-in was at 8AM. We were about 25 miles east of the facility with a four lane divided highway connecting us to the base. As on-base housing was tight, many of the younger officers were forced to live in town. To make housing problems worse, we competed with poor starving college students. We were all almost the same age and many friends were made between the two groups.

I generally rode with my crow (EW) to base as it was far easier to get a car ready for the drive home with two people working to get it unfrozen and started. It also freed up my car to allow the wife an escape to town so she could find a parking space to plug the engine heater in to. If you were really lucky you could get inside heated parking to defrost your car. It was a lot like a freezer inside your refrigerator. There was always a race for the big airplane heaters on the flight line to melt the ice and get the doors open. It was not unusual to cook the paint right off the vehicles. Oh what a challenge, and rank was not an issue in that endeavor. Some young airman would make a few extra dollars from the cash-rich officers and they always seemed to control the heaters.

Once we were called early to report for alert and I called my crow, who advised me that there was no way was he going to chance taking his new Pontiac Gran Prix convertible and just perhaps if we waited the changeover would be delayed. Well then the radio announced that no civilians need to report, no non-essential military were to report, and finally the announcement that only active duty alert crews were to attempt the drive in the howling blizzard for the mandatory 8 AM crew change. John showed up ready to go since he certainly did not want to test my VW for the task. High winds were known to blow VWs sideways and right off the road. Now you'll understand why all the winter weather gear was issued in August for use in November through May.

We started out, with the lights on, and driving very slowly while looking to the side of the road for the freeze or plow lines. It was important to always carry a 50-foot piece of nylon cord in your pocket to anchor you to your car while you installed warning flags so the plows would not grind you up and allow a positive way to return to where you started. It did not take long to see other folks on the side or middle of the road with flares hoping to catch a ride in from their abandon vehicles. Well it was quite a trip and in the end we had 16 folks in the car and I assure we were cozy. As we approached the main gate the announcement came that the base was closed and on a stand-down mode to protect valuable personnel

It so happened the sky cops, with their tracked vehicles, were directly across from the O'Club. We delivered our busload at 11:15 and got help getting to the club for lunch and libation. Only the night duty officer was there and apparently there was an open bar. We rewarded ourselves for following orders to report at great risk by making cold sandwiches in the kitchen. We had done considerable humanitarian work with fellow lost souls and repeating the SAC Motto that "Peace Was our Profession."

We got carried out of the club to the alert shack covertly. We made the crew changeover on time at 10PM and slept the night away peacefully.

The weapon leaves the airplane at 400', 325 knots indicated airspeed.

Close Counts in Nukes Too...

Dave Volker

In June of 1973 I brought home some news that did not make my wife of about three years very happy. I told her that I had just been assigned to a TDY that would take me away from home during our third wedding anniversary. Even though we had been married almost 36 months I had spent less than half of it with her. As a copilot of the 46th Bomb Squadron at Grand Forks AFB, North Dakota, I served a regular nuclear alert schedule for the first 24 months of our marriage. That was interrupted by a 179-day Arc Light TDY.

After returning from that I was given the task of being the 319th BW Air Weapons Officer. I was assigned an office in the alert facility and sent to Carswell AFB to a six-week intensive training program on how to be a Nuclear Weapons Safety Officer. During that short time I learned more than I will ever need to know about the internal workings of both the conventional and unconventional weapons we could deliver from the B-52. I also learned how to convey this vast store of marginally useful knowledge to crewmembers during training sessions that were mandated by the recurrent schedule and loathed by all. How I got this truly plum assignment is a story in itself that I won't go into right now. Let's just say that Col Lee R. Senter, the Wing Commander, felt the need to punish me for a non-flying infraction, and this was his way to do so.

My July 1973 TDY assignment was to be the official DOD/Air Force observer for a series of nuclear weapons tests to be conducted at the Tonopah Test Range in Nevada. It seems that the weapons that our H-models were tasked with delivering needed testing.

The nuclear test ban treaty we had with the Soviets prohibited the full scale testing of nuclear devices but it did not prohibit the testing of the non-nuclear components. To prepare for this series of tests, two fully operational B-28FI model nuclear gravity bombs were removed at random from our inventory at Grand Forks by Sandia Labs representatives. They were then transported to the Pantex factory in Amarillo, Texas, where under careful guard, all of the nuclear and high explosive components were removed. The space created inside, by the removal of these parts, was filled with equivalent weights and telemetry gear. A small flash charge triggered by the firing circuit was fitted to an opening in the casing to better note the exact time of detonation. If all went well, when these weapons were released under simulated EWO conditions from a B-52 bomb bay, their highly complex and very reliable internal components would spring to life and perform their necessary work to fuse and fire the weapon. The telemetry equipment would report the internal workings as the components prepared to create a small sun for a fraction of a second near the point of impact. The results would hopefully tell the folks at Sandia Labs useful information about how well these devices aged in storage. The nuclear and explosive components were assumed to be very reliable. Could we trust the other parts after they had sat dormant for years awaiting what might be the end of the world? Was the sword still sharp, even after leaving it in the scabbard all this time?

For my part, I traveled from Grand Forks to Las Vegas via Frontier Airlines. That was a real luxury as on my previous TDYs I flew almost as cargo in the back of KC-135s. Such was the life of a crewdog. I was afforded the ride in a civilian airliner probably because the T-29 liaison aircraft never went to Nellis. In 'Vegas I rented a car and drove Hwy 95, for 215 miles north and a bit west to Tonopah, Nevada. That was the era of no posted speed limits. I found out that a Chevy Impala could easily cruise at almost 100 mph. It was rather like low level flight, it seemed.

After a memorable night in Tonopah's Silver Dollar Hotel I got up at 0: Dark 30 and dressed in my summer blues and departed for the test range. The range was over 30 miles outside of town and I had to be there just about sunrise. The test would be conducted about 0600, after sunrise but before the heating of the day made long range photography difficult. I drove there over some very lonely roads, checked into the range observation post and stood waiting for the show to begin.

In the meantime, the weapons that had been removed from GFAFB inventory months before had been shipped back to our base and on the previous day, loaded onto one of our line birds. Bomb-Nav had prepped the BNS for as much accuracy as could be achieved. The squadron's best radar nav was pulled off of alert duty and assigned to a "5X" crew specially constituted for this test. This high time captain RN had a notoriously bad attitude but they could not have chosen a better person for the job. He consistently got the best RBS scores even if he had the most abrasive sense of humor and wrinkled uniform the wing brass had ever seen. In the right window seat, sat one of the most "SACumcised" pilots I have ever flown with. I swear this guy stayed awake memorizing his T.O. B-52H Dash-One while his wife slept beside him. He not only understood the aircraft intimately, he could also fly very well. He could be relied upon to drive exactly where the RN wanted him to in strict accord with operational policy and doctrine.

They had departed the Forks the previous day, taking off at about 2100 hours local. Their mission profile was to simulate that of an Emergency War Order mission as far as the weapons were concerned. They first climbed to altitude and refueled over eastern Montana. They then climbed further to cold soak the weapons. They had to avoid heavily populated areas because they were carrying what could be mistaken for the real thing. In the lower two shackles of the four-weapon clip hung the two modified B28s. In the upper two shackles, two "shapes" were hung. These were B28 casings that looked and weighed the same as the lower two but contained only parachutes and the necessary mechanisms to deploy those 'chutes once released. These were meant to test the ballistics tables for the B28 when they were dropped.

Around and around in a high altitude racetrack over northern Nevada they spent their late evening and early morning. Just as dawn was about to break over the hills to the east, they began their descent - down the long slide to about 200 ft. AGL and 325 KIA. It was as close to a real nuclear bomb run as any of us, either on the ground or in the airplane, would ever want to experience. The RN locked onto the radar reflector placed immediately adjacent to the center of the target. Even from 200' it stood out like a beacon. He had to turn the gain way down, just to find the center of it. All other returns faded away. Initial Point inbound, he gave the command to center the Pilot's Directional Indicator.

I had heard them call IP inbound over the range frequency. I was about half a mile from the target and from my slightly raised vantage point in the range tower I could see the radar reflector in the distance. I had never seen a low level B-52 approach a target at bomb run speed but I was about to. I turned to my left and out of the corner of my eye I saw a flicker against the desert background. I called it and all eyes in the tower turned where I was pointing. Hugging the valley floor, a seeming ghost with white belly and camo top flashed between the scrub and small hills. Then the nose rose out of the brush briefly and leveled off. The popup maneuver was executed perfectly. At the same time, the bomb bay doors slammed open and a small cylindrical shape emerged from the belly of the plane. Almost instantly gravity began pulling it to earth and the steel plate that had held the white nylon parachute incased all these years separated. Amazingly, the parachute deployed while the bomb was still almost horizontal to the ground. By that time the bomb bay doors had cycled shut and the aircraft was descending back to 200' above ground level, getting separation from what would have become a 1.5 megaton ground burst, had this been for real. One swing back and about one half of one swing forward was all the weapon had time for before it slammed into the target area.

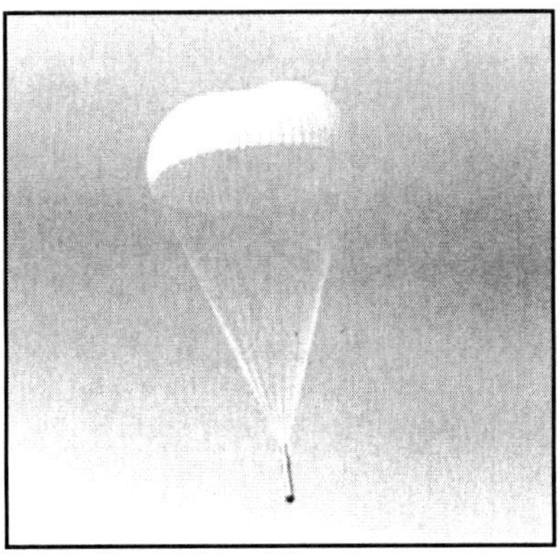

The parachute has deployed and the weapon nears the target.

Then I heard the range officer as he looked through his high power binoculars at the weapon and the target say, Oh My God!"

I didn't know exactly what happened that concerned him but everything seemed to be OK. The approach looked good, the delivery was text book, the parachute had deployed ... what had gone wrong? There was a moment of silence as we all waited for the internal mechanism of the weapon to work. I knew this to be a bit over 90 seconds so I asked what was wrong. The answer was not what I had expected.

A small flash charge notes the moment that the weapon would have detonated.

"They almost hit the target!" was the response from the amazed Sandia Labs employee.

My first non-verbal reaction was "I thought that was the point!" What I found out later was that the weapon had almost struck the radar reflector and center point marker placed at the very center of the target circle. The weapon came to rest only 57' from the center of the target. It was the most accurate delivery of a parachute retarded weapon at the Tonopah test range to date.

The weapon came to rest 57 feet from the target center pole. The radar reflector is nearby to the left.

They made three more passes at the target and successfully released the second weapon and the two shapes, one on each run. The average error for the four drops was less than 200'. On the final run the gunner radioed me to inquire about whether I had a visual on the aircraft following them. It turned out to be a flight of two F-4s from Nellis. It seemed they hadn't been told that a B-52 would be using the range that day and decided to pursue. The range officer told them to break off and go home. The show was over.

Close counts in horse shoes, hand grenades, and nukes too!

What a day! What a performance by an outstanding crew and their aircraft! I was privileged to have witnessed it all and have the pictures to prove it. Thanks Col Lee R. Senter, for punishing me in that manner

Chapter Three - Emergency

Emergency - [ih-mur-juhn-see] – *noun* - a state, or need for help, created by some unexpected event.

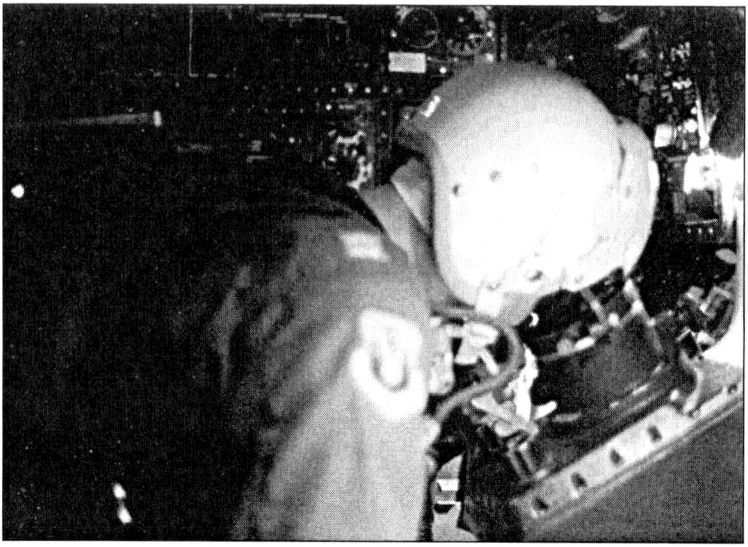

Gene Daspit in a downward ejection seat looking through the optics of a "D" model.

Beating the Odds - Escape from Swoon 52

Gene Daspit

As crewdogs, we woke up every day with a certain set of unquestionable beliefs – things we knew to be true: No one wins in a land war in Asia ... no battle plan survives contact with the enemy ... no one survives a B-52 downward ejection at low level .... I am living proof that sometimes conventional wisdom is wrong.

After a four year tour at SAC Headquarters, I returned to flying as a Radar Navigator assigned to the 92nd Bomb Wing, Fairchild AFB, WA. After a short term on a line crew, I went to work for Bomb/Nav as a Target Study Officer. I was called in on Tuesday, October 16, 1984, to substitute for Maj Mel Duarte, Radar Navigator, on crew S-02. This Standardization and Evaluation (Stan/Eval) crew consisted of Capt Robert L. Kenney, Aircraft Commander, 1Lt Douglas J. Schwartz, Copilot, 1Lt Kenneth O. Portis, Navigator, Capt Sean M. Yeronick, Electronic Warfare Officer, and Sgt David W. Felix, Gunner. Col William L. Ivy, the 92nd Bomb Wing Deputy Commander for Operations, was on board as the safety observer. We were flying B-52G 57-6479 with the call sign Swoon 52. The first part of the mission called for an air refueling followed by low level night terrain avoidance training with three bomb runs at Holbrook, Arizona.

Takeoff through air refueling was uneventful. After air refueling, I ate my lunch which my wife had made for me. Part of my lunch was a slice of homemade peach pie which I decided to eat instead of having it after low level. (I know now why so many people say "eat your desert first!") Even though I had never flown with this crew before, I was impressed by how well they flew together. I had flown previously with the EW, Sean Yeronick, whom I had crewed with prior to my move to Bomb/Nav. They were sharp as a Stan/Eval crew should be and I felt comfortable with how professionally they approached their duties. As an old crewdog you get a sense for which crews work well together and which ones don't. These guys were good and I fit right in with them.

Our entry into Holbrook was on time and on course and we were descending down to our initial low level altitude over Monument Valley, Utah in Arizona, prior to our terrain avoidance calibration area. Monument Valley is an area with the massive red-orange buttes rising 500 feet above the valley floor that you see in the commercials and many movies including John Wayne movies. Doug Schwartz, the copilot, was flying the penetration into low level when he reported his radar altimeter didn't appear to be going down. I immediately checked the radar altimeter ribbon on my radar screen which indicated we were above 2,000 feet above the ground and descending. At that point, I heard Bob Keeney announce over the interphone that his radar altimeter appeared to be normal and I then confirmed that my radar altimeter agreed with the pilot's. I then looked at my low-light television monitor/Electro-optical Viewing System (EVS) and saw we were heading directly at a mesa which I later learned was called Hunt's Mesa. Normally terrain features have a slightly out of focus – 'washed' appearance. I was surprised to be able to make out individual rocks and boulders on the mesa and no horizon. It was a little too clear. I then told the pilots to climb but they never heard me because both the pilot and copilot saw the mesa and started an immediate hard climb to clear it.

As we started climbing, I saw the mesa drop off the bottom of the EVS screen. I then felt the plane shudder a little like we had hit some turbulence. It felt like it came from the tail of the plane but I had felt B-52s shake a lot worse when hitting large birds or having surface-to-air missiles go off in close proximity over Hanoi. I didn't feel any concern although for some strange reason I thought to myself, "Oh, we're still flying". All of that happened in a very short time of what seemed to be about 10 to 15 seconds at the most. I was still sitting watching the EVS feeling comfortable about the situation, aware of normal cockpit lighting around me when I glanced over at Ken Portis. In the instant it took for me to turn my head to look at him, all the cockpit lights had gone out and I could only see him in the bright glow of the "Bailout Light" shining on his helmet. He already had his hands on the trigger ring to start his ejection from the aircraft.

After that, what happened was a reflex action because I have no recollection after seeing him in the "Bailout Light" of thinking that I had to pull my trigger ring and eject from the airplane. Ken already reaching for his trigger ring to eject and as it turned out, I actually ejected from the airplane before he did. The next thing I remember was a bright orange light in front of me like a bright light that was so intense I was seeing it through closed eye lids and it puzzled me. (Evidently it must have been caused by an explosion because I had some 2nd degree burns on my hands and face, my parachute had burn holes in it, and the navigator had bad burns on his face, hand, and the back of his neck.) Next I felt my helmet come off my head, then in the darkness I could just barely make out the shape of the plane with just a little flame coming from it. I also recall about that time feeling the opening shock of my parachute. I then saw another parachute and was watching it in the light of a second fire ball from the plane when I decided I had better stop sightseeing and get ready to land. I couldn't see the ground and was trying to figure out where it was. I got my feet and knees together with knees bent and pointed my toes down like we had been told to do in survival training when I hit. As I hit, it looked to me that I was five to ten feet from the edge of a cliff. We had landed on the top of the mesa that we had hit and as it turned out the cliff at the edge of the mesa right in front of me dropped off about 500 feet almost straight down.

The shudder that I had felt was the right wing tip hitting part of the mesa as we flew up through a small saddle in an almost sheer vertical wall. The mesa tore off twenty feet of the right wing and the two engines in the outboard pod on that side. To put things in some sort of time perspective, from the time the wing hit the mesa until the plane finally impacted the ground was about eight seconds and was a distance of about 4300 feet. Part of the eight seconds was while I thought everything was OK. I landed about 1800 feet from the airplane's final impact point, hitting the ground about the same time it did. Everything I just described above actually took place in only around three to four seconds!

There were five survivors out of a crew of seven on the aircraft. Colonel Ivy was in the jump seat and did not have a chance for a manual bailout. The gunner, Sgt Dave Felix, ejected but died on impact with the ground after his parachute was damaged by debris and did not deploy. We were told we were the first survivors of a B-52 crash while flying a low level training mission. Fortunately, there had been a modification to the navigators' downward ejection seats that lowered the minimum altitude for bailout to 250 feet with everything working automatically and properly. I was told I ejected well below the 250 foot minimum with estimates as low as 138 feet above the ground and that my parachute deployed about fifty feet above the ground.

As I hit the ground, I hit on my feet and then went down on my side losing my breath on impact. I felt like I was a sack of potatoes being dumped off the back of a moving truck. I also knew instantly that I had broken both legs and it was bad. The first thing I did was to unfasten my parachute so it wouldn't drag me across the ground since I could see it was still billowing in the wind. I then tried to get as comfortable as possible and pulled in my survival kit to find something to keep me warm and try not to go into shock as it was cold and windy with a chill factor of about zero degrees. While I was still doing that, I heard Sean Yeronick, the Electronic Warfare Officer, calling to find out if anyone else was around. I called back and told him I had broken both my legs. He told me, he and Bob Keeney, the AC, would be there shortly. At that point, I became concerned about them wandering about in the dark trying to find me with the cliff so close to me. I knew if they fell off the cliff, they wouldn't be any help to me so I told them about the cliff. They told me to call out so they could find me but I was still concerned. I then remembered there was a strobe light in my survival kit so I reached in, found it, and turned it on. About five minutes later, Bob showed up followed a short time later by Sean who had a badly sprained ankle from his parachute landing. Bob's only injury was "whiplash" caused by still being bent over reaching for the bailout light switch when he initiated his bailout sequence. When Bob got there, he took a look at my legs but didn't say anything. Sean seemed to me to be shaken up so I had him lay up against me so we could try to keep warm. Bob then pulled my parachute over the two of us and I was surprised at how much warmth it provided us. With my broken legs, there was no way I was going to attempt to get into the sleeping bag in the survival kit. I did get it out and tried to just lay it over me to get a little protection from it. From where we were, we could see the town of Kayenta, Arizona, which was about ten miles away. We could see cars streaming out of town in our direction and figured it would only be a matter of time before help arrived. Bob then took the strobe light and tied it on to a small scrub tree right in front of us at the edge of the mesa. I also figured there would probably be another B-52 flying over shortly.

About 45 minutes after we crashed, a BUFF from Ellsworth AFB, South Dakota, flying Holbrook, flew over and flashed his lights but kept on going. It was also around that time, a vehicle started to approach us. It appeared to only be a couple hundred yards away, but was past the cliff. Bob went over to the edge of the cliff and heard someone calling asking if anyone was up there. He called back and the voice told him to stay where we were because there was a ravine in between us. (That was an under-statement. It was a 500-foot deep canyon and we weren't about to go anywhere in the condition we were in, especially me.) He said he would be up to where we were in a little while. It took him over an hour to finally reach us. In the meantime we heard another voice and the pilot went off looking for the source. He came back and told us he had found Ken Portis, the nav Ken was down in some rocks about 30 yards from me but he couldn't move because of injuries. He had landed in some large boulders and had torn up his left leg and left elbow. Ken laid in that location until the rescue helicopter came to pick us up. Bob went back to help Ken as best he could.

When the vehicle that we had seen earlier drove up, it turned out to be a Navaho tribal policeman and a Navaho shepherd in a four-wheel drive vehicle. The area was only accessible by jeep and other four-wheel drive vehicles. I figured then I'd be able to get some sort of medicine for the pain. The pain I was feeling was at times dull and steady and a few other times unbearably excruciating. Lying on the freezing desert floor probably didn't help the cause either. I had taken some aspirin from the survival kit – not only was it ineffective for the pain – I found out later that one should never take aspirin if experiencing any kind of bleeding – and I had quite a lot of that with my injuries. I asked him if he had anything but he said he didn't. They decided to build a fire near us to help us stay warm so they found some brush to start a fire but they didn't have a match or lighter so I again went digging in my survival kit and found matches for the policeman. He used them and some paper from his clipboard to start the fire.

Eventually the B-52 came back over and the crew was able to make radio contact with Bob Kenney. Ken Portis had just gone through survival training a couple months before and remembered all the details and frequencies for the survival radio so he was able to remind Bob of them, facilitating communications with the BUFF that was overhead. Bob vectored the BUFF in and gave a countdown just like we were trained to do in survival school. By that time the weather started to get real bad, windy, low clouds, and snow flurries. Earlier the policeman said there was an Arizona Public Safety helicopter on its way but then he told us it wasn't going to be able to get there because of bad weather between us and where the helicopter was coming from. That was the first time my morale got down a little. It was about that time a lot more people started showing up. I realized that Sean was still lying up against me trying to keep warm, so I told him to go get in the four-wheel drive police vehicle that was just sitting a short distance away with the motor running and lights on. I told him there was no use in two us laying there in the cold and that at least he would able to get warm, so he got in the vehicle.

By then we had been on the ground about four hours. One of the people who came to check on me was a short little Navaho woman. She told me she was an EMT and started checking me over. I told her my legs were broken but my back was OK, then I asked her my often repeated question of the night, namely did she have any pain medication, to which she gave the same answer that she didn't. She then asked someone to get splints. I found out later the person helping her was the 16 year old son of a volunteer fireman from Kaycnta and had a very good knowledge of first aid since he was an Eagle Scout. They couldn't find any suitable wood for a splint but he remembered there was a barbeque grill in the back of one of the jeeps they had driven up to the crash site. He helped the EMT splint my legs which was quite painful. About 15 minutes later the doctor from the Public Health Clinic for the Indians in Kayenta showed up. He and the EMT from the clinic started out in their ambulance but got bogged down in sand. The light from the explosions from the crash were seen up to 60 miles away. When calls started coming in, the Kayenta Volunteer Fire Department started out in their personal jeeps to see what they could do to help. They were familiar with the areas as they regularly camped out and held cook outs up on Hunt's Mesa. As they were heading out to the crash site, they picked up the EMT and the doctor and drove them up to the top of the mesa. When the EMT told the doctor I had multiple compound fractures of both legs, I knew for first time how bad they were broken. They had both broken just above the boot tops with my heels bent all the way back touching the back of my thighs and they were twisted around each other. When I later visited the classroom of one of the volunteer fire persons from Kayenta, her fourth grade class called me the "human pretzel". I asked the doctor if he had any pain medication but he said no. I found out later that he was well equipped if any of us had had a heart attack. He then told me he was going to straighten my legs which would help my recovery. Once again I had an excruciating shot of pain to go along with the constant dull pain I had and was able to sort of tolerate.

During this time the other BUFF continued holding station above us – at about 200 feet above the mesas and in increasingly bad weather. I was a little concerned that they might hit a mesa themselves. Bob, who had been talking to them, came over and told me there was an Air Force HC-130 "Hercules" and an Air Force HH-53 "Super Jolly Green Giant" helicopter from Kirkland AFB in Albuquerque, New Mexico, on their way. They were instructor crews from the combat crew training squadron at Kirkland. A short time later the HC-130 came over and started dropping flares. The scene seemed like it was out of a movie, almost unreal and it was the first time with the orange light of the flares that I could see how rugged the terrain was, the mesas around us and the deep canyons with the cliff right in front of me.

By that time there were all kinds of people around but nobody seemed to be doing very much. Occasionally, someone would ask how I was doing and I told them I was doing fine. I wasn't aware at the time that they were busy looking for other survivors besides us four on the Mesa. The Kayenta Volunteer Fire Department coordinated and led most of the search of the surrounding area. At some point Bob came over to me and told me they found a body, which at the time he thought was the copilot, Doug Schwartz. Several Navahos later found Doug alive only a couple hundred feet from where the plane hit. The body they found was Dave Felix, the gunner. Silently we all knew that Colonel Ivy never had a chance to get out of the aircraft – not being in an ejection seat, he simply didn't have enough time at that altitude to make a successful manual bailout. It was several days before the Air Force finally confirmed what we already surmised.

In their efforts to try and keep me warm, several of the Navaho volunteers built a second fire on the other side of me. I started to get concerned as the wind began fanning the flames showering sparks down on the parachute covering me. Being strapped down on a barbeque grill between two large fires made me wonder for a moment if my final fate was to be the main course! It seemed funny to me at the time and after several hours on the ground, I had to do something to keep my spirits light. A number of people have since commented to me about their images of me being on a barbeque grill between a couple fires and surrounded by a bunch of Navaho Indians.

About six hours after the crash, the Air Force helicopter arrived, landing about a hundred feet away. The paramedic came over to check on me. He had already been informed of my injuries and so I was sure he was bringing some pain medication. It was the first thing I asked him and he promptly told me no. What? A Medevac helicopter without medicine? I found out later, they had been on a training mission that night and in the rush to get to us, had forgotten to get their medicine kit. I also learned the HC-130 was in such a hurry to get off the ground, they had to get flares off the other planes on the ramp rather than waiting to get them out of storage. I told him I was doing fine. He then got some air splints and they put those on my legs and put me in a heavy duty sleeping bag on a litter. When they did that, I got another bad shot of pain. After attending to me as best he could, the paramedic and the helicopter crew lifted off and moved over to where the copilot was. Fearing he had serious head injuries, no one made any attempt to move him over the ground to join us. Doug had been the last to eject from the aircraft and was a considerable distance from the rest of us. He had split his helmet open on landing and had remained unconscious for some time, not remembering anything after initiating his ejection. His first memories were of the Navahos who found him hours after the crash. The paramedics attended him, loaded him aboard and returned to pick the rest of us up. Doug's only other injury was a broken shoulder. The helicopter crew reported that their Inertial Navigation System (INS) showed they landed only about one hundred feet from the coordinates the Ellsworth BUFF had relayed to them.

Being aboard the helicopter seemed surreal: a warm place with light! It felt a lot better than our previous hours in the cold and wind. The inside of the helicopter looked huge. It seemed like it was the size of a HC-130 so later when I got another look at one, I was surprised it wasn't nearly as big as I thought it was that night. I guess the perspective of your environment is different when you are flat on your back on a medical stretcher! We finally took off about eight hours after the crash. It was about an hour flight to Farmington, New Mexico, which had the nearest suitable medical facilities. During the flight, the paramedic came back to check on me and to apologize for the rough ride caused by the turbulence and again while the helicopter was mid-air refueling with the HC-130, although I thought it was quit smooth. I guess he had never flown in a BUFF at low level over Kansas in the middle of a hot summer day.

When they finally got me into the Emergency Room at Farmington, New Mexico, I knew that finally someone would have something for the pain! I asked one of the nurses who told me they had to get a urine sample first. That wasn't as simple as it seemed. They stood there trying to figure out how to get me out of my flight suit and long johns without causing me a lot of pain. To me it was no problem. I informed them they could just cut everything off. I wasn't worried about that since the only thing I was wearing that was mine was my undershirt and shorts. The flight suit, long johns, socks and boots were all Government Issue and their destruction was very minor compared to the value of a B-52. They got my clothes off and once again I told them they could do the same thing to my boots and socks. I remembered lying there being aware of a new nurse coming in to help. She started to roll down the sock on my left leg and I promptly told her to just cut it off.

They finally got everything off me and a nurse came to get the urine sample. After waiting for hours for relief from the intense pain, with just one simple task to perform before they could administer it, I couldn't go! She must have stood there about 10 minutes waiting for me. It was about that time I became aware of water running in a basin nearby. I asked her what that was for. She told me it was to try to get me to go but the only effect was that she now had to go … bad!. I realized that it would probably take me a couple hours so I told them to use a catheter. That gave me another jolt of pain, but shortly after that I was finally able to get some pain medication.

By that time, the hospital was a zoo with the staff doing their best to take care of the five of us, the local TV and news people trying to get whatever information, photos and other tidbits they could get and the remainder of the staff trying to keep them away. My 15 minutes of fame came when a cameraman snuck a picture of me being wheeled into X-Ray. I made the ABC Evening News! As I came out of X-Ray, the staff was able to get my wife, Mavourneen, on the phone back in Spokane. She asked me how I was and I gave my now famous statement: "I am fine – I only have two broken legs"! She wondered how many legs I had! I knew my condition could have been a lot worse. The pain in my legs was good because it meant I hadn't broken my back. Fortunately, breaking my legs absorbed the force of the impact and saved me from a permanent back injury. She also asked me if I had eaten that slice of peach pie. I assured her I had.

After that, they wheeled me in to the operating room to set my legs. They gave me a general anesthesia and a spinal. I remember waking up about an hour later while they were still operating on me and talking with the anesthesiologist for the next hour before drifting off again. Later, when I woke up in my room, I discovered what they had done. My left leg was in a half-cast and wrapped with an Ace bandage. My right leg was something else. I had ten steel rods sticking out with everything being held together by a steel apparatus that could best be described as looking like some sort of erector set. The medical term for the apparatus was a Hoffman External Fixation Device. I was glad it was winter and I didn't have to worry about any thunderstorms with all that metal sticking out of me.

That was about the time things really got fuzzy and weird. We expected to be in the hospital in Farmington for maybe about a week. We heard that the Bomb Wing Commander was going to fly our families down to Farmington on a KC-135 tanker and fly us back. By Thursday afternoon, we were told that we were to be flown back to Spokane on an Air Force C-9 Medevac plane. This was about 48 hours after the crash. That evening I remember the hospital staff putting me in an ambulance in Farmington along with the copilot, Doug Schwartz. It was the first time we had spoken since before the crash. The next thing I remember was waking up and seeing my wife and a friend who was one of the chaplains at Fairchild. After saying "Hi" to him I asked what day it was. They told me. I had been in a coma for those five days.

Evidently, in flight on the C-9, I had passed a fatty embolism from the area of the fractures into the blood stream and it went to my lungs causing me to have a lack of oxygen. When we arrived at Fairchild AFB, the flight surgeon knew right away what the problem was and got me stabilized and transferred to a major civilian hospital in Spokane. Talking to him later, he said the measurement of oxygen in my blood indicated that on the flight I was past where brain damage should occur and was down where I probably should have gone into cardiac arrest.

When they got me into the hospital in Spokane, I was put in the Intensive Care Unit and was on a respirator. I was only on the respirator for about 24 hours, a lot shorter time than the doctors had predicted. During that time one of the more amusing episodes occurred. Visits to the ICU were limited to ten minutes. I had been in for about a day when my wife and my mother were on their way to see me during one of those ten minutes visits. They heard this loud shrieking and screaming coming from the ICU. They were met by one of the nurses who told them that they were trying to change an IV and I wouldn't let them, and I was yelling that they were trying to kill me. The nurses had noticed that whenever my wife was there, I would calm down so they thought maybe she could calm me down. As she leaned over me, I lunged up and bit her on the neck, setting everyone off in a panic except my wife. All she could do was laugh and ask them what they had been feeding me. As it turned out, they had given me a blood transfusion and it had been a full moon the night before. She still has a scar on her neck.

The doctors had me doing all kinds of therapy, especially for the brain damage they thought I had. During my stay in the hospital, the internal pin in my left leg slipped out. I went in for another surgery and the pin was replaced with another Hoffman External Fixation Device. My legs then looked like a major construction site with all the metal sticking out of them. As bad as it looked, I was able to get up and walk around with a walker. I spent two months in the hospital, finally getting home about five days before Christmas. I was at home for about two and a half months with the external devices before I went back in for a short stay surgery to have them removed. The Orthopedic Surgeon then gave me plastic braces. I was able to get rid of the left one about the beginning of June and the right one about the beginning of August. The doctor finally let me start running about the beginning of October. While we were on Hunt's Mesa, Sean Yeronick said that I probably wouldn't be able to run Bloomsday. Bloomsday is a 12 kilometer race in Spokane held the first Sunday in May and at the time was the second largest timed run in the country with about 38,000 participants. I told Sean I was planning to run it, so it became one of my goals for recuperation. I wasn't able to run it in 1985 but did complete it walking about a third of the course and riding my wheel chair. The following year I did run it finishing with my fastest time up to then. There is a saying that "any landing you can walk away from is a good landing". I guess for me it was a "good landing" even though it took me almost a year before I was able to walk away from it.

Later, both Bob Keeney and I went through a Flying Evaluation Board and were cleared and returned to flying along with the rest of the survivors. As with most accidents, changes came about. Of the five survivors, three of us lost our helmets during ejection even though they were properly fastened. Fortunately, the two who didn't lose their helmets were the copilot and navigator. They both came down in rough terrain with the copilot's helmet splitting open on impact. The strap on my helmet, which I still have, tore away from the grommet that held it to the helmet. As a result, B-52 crews changed to a light-weight fighter pilot type helmet. Also, charts for low level navigation were changed from ONC charts to TPC charts to show the terrain in better detail.

In 1986, after everything was finished, the SAC Chief of Safety asked us if we would do a training film. In February, the five of us who survived, went to Carswell AFB, Texas, to make the film. Since we were the first to survive a ground impact during low level flight training, we were interviewed about our decision to bail out. The SAC Chief of Safety felt we had a lot to offer since he thought that in many previous aircraft accidents the decision to bail out was delayed until it was too late to safely eject. While we were at Carswell filming, we went to get some shots inside a BUFF that had the pilot's hatches removed. Doug Schwartz is a big guy and when he stood up on the seat, he didn't make it through the hatch. He was measured and his shoulders were about a half inch wider than the hatch opening, accounting for his broken shoulder. In the process of making the film, there was so much said about what we did in a survival situation, that a couple months later, the crew went to Kirkland AFB and was flown on helicopters to Hunt's Mesa to shoot some on site footage with each of us at the location where we made our parachute landing. We flew out to the mesa in several Hueys. While flying out to the crash site, we encountered a thunderstorm and had to land out in Monument Valley. Naturally, I got very apprehensive being near a thunderstorm in a flying machine, especially a small one the size of a Huey.

The next day after we had finished filming, Bob Keeney, Doug Schwartz, Ken Portis and I drove out to Kayenta to meet some of the local people who assisted us that night. That evening while we were meeting a lot of the people, I was given a periscope sextant that one of the members of the fire department had retrieved from the wreckage. The Assistant Fire Chief for the Kayenta Volunteer Fire was Debbie Braff, a fourth grade teacher at Kayenta Elementary School. She had previously inquired about how we were doing after the accident and then sent an account of what she had witnessed that night to Doug Schwartz as he did not remember anything that happened for several hours. The next day, we visited Debbie's class and talked with the students who named me the "human pretzel". I also found they were using a radar film magazine for a doorstop for the class room.

That following summer I took leave and traveled back to Kayenta with my family. We stayed with Debbie for several days and to this day, we still keep in touch. While we were there, she and several other members of the fire department took us up to the crash site in their jeeps. My two teenage children and my wife had a great time. I was scared to death by the trails they took us over to get up to the top of the mesa. I was hanging on to the jeep I was in with white knuckles. This was the third time I had visited Hunt's Mesa, once ejecting out of a BUFF, once flying up in a Huey, and this time in a jeep. I have to say, the jeep was the scariest and next was in the Huey with thunderstorms around. The least scary was ejecting at extremely low altitude. It happened so fast, there was no time to be scared. I owe my life to the training and thousands of hours of experience that when the time came to eject, it was a pure reflex action on my part. If I had taken time to analyze the situation, I probably wouldn't be writing this.

On a final note, when something like the crash happens, rumors spread throughout the command. I couple years later I was attending a nuclear weapons class at Kirkland AFB. One of the other members of the class was John Martin who was the Wing Navigation Officer on Guam while I was the Wing Target Study Officer. He told me he had heard a rumor that I had suffered brain damage (I assume the brain trauma from the embolism) because I had landed on my head. He said he knew that wasn't true because I was so hard headed that if I had landed on my head, nothing would have happened. Like we always said, "you have to be tough to fly the heavies".

B-52G 57-6468 at the Kenney Gate of Offutt AFB, Nebraska.

Bird Strike at 440 Knots

Gelvin Nicely

Unlike their commissioned officer colleagues, enlisted crewdogs couldn't avoid completing the essential requirements to stay on active flying status, even when they were assigned to Strategic Air Command Headquarters. It really made sense to stay current in one's flying profession so that our "heads were on straight" when we made decisions at the office that would directly affect the crewdogs in the field.

Of course, Offutt AFB, Nebraska, didn't have any B-52 aircraft assigned, so I stayed at least minimally current as a B-52 gunner by traveling to various bases several times a year to fly a few sorties. I actually flew quite a bit during 1987 in both G and H models. Feeling no need for a practice mission with a local instructor, I scheduled my annual flight evaluation mission as a single sortie from Mather AFB, California. I figured it would give me plenty of time to complete written tests, publications checks, and mission planning, and if needed, to schedule another sortie to complete my check ride.

I was assigned to crew S-03, a 320th Bomb Wing Standardization / Evaluation crew. My evaluator was their gunner, SSgt Ken Baily, who I had helped train and evaluate while we were both assigned to Ellsworth AFB, South Dakota, a decade before. Our planned mission was to be a three-ship cell departure, followed by inflight refueling with two tankers, then low altitude navigation and simulated bombing using an in-trail formation to simulate conventional (as opposed to nuclear) tactics, then back to Mather for some practice landings. On the radio we would be Ursa-03.

Completing all my preliminaries uneventfully, I met my crew about 0615 on 4 November 1987 to get life support equipment, box lunches, enroute weather forecast, flight plan filing, and preflight inspections done in time for our leisurely 0900 takeoff time. Ken squeezed into the nook behind the EW station that served as the defense instructor position. Amazingly, our three B-52s got airborne on time and we did the usual equipment checks while we controlled timing to our air refueling date with the tankers.

As I recall, the refueling was pretty routine, and soon we were on our way north and east to our low level instrument route, IR-473, which passed through eastern Montana and western North Dakota. We left the cell lead position for number two because of some training considerations. My fire control system was in good shape and I completed my "on watch" procedures before arrival at the simulated threat area. I remember feeling pretty good about how the check ride was going—just had to monitor bomber number three a quarter-mile behind us and watch out for nonexistent "bogies."

Our navigator re-briefed the entry procedures for our low-level route and we entered the route on time in cell formation and cleared off the air traffic center radio frequency. We descended to about 500 feet above ground level, calibrated the terrain-avoidance system, and got the speed up to around 440 knots. Just another hour or so to the target area, then climb out of the route and head back to home base. The weather was clear with a little choppiness, but it was a great day to be flying.

And then it happened. With no warning there was a loud BANG followed by a strange, loud roar. Instinctively, I looked up and noticed the "bailout light" on the panel in front of me was not lit, meaning none of the six ejection seats had been fired. The EW and I exchanged quick glances as we unthinkingly hooked up our oxygen masks and assumed the right position to eject from the airplane. A look over my left shoulder and the seat-of-the-pants feel told me we were climbing and I saw the pilot pull all the throttles back. Ken and I exchanged "thumbs up" signals as one of the pilots told us what we needed to know: we had taken a hit in the nose radome by a large bird, and we were slowing down and climbing out of the route.

What we couldn't see, but found out later, was the bird had knocked a significant hole in the radome which allowed that 440 knot wind to come inside it. That was the cause of the bang and the roar. The pilots later told us that the radome immediately swelled up like a balloon and they feared we would lose it if we couldn't slow down quickly. Among other things, losing the nose radome of a B-52 in flight also means losing all airspeed instruments—not a fun situation!

We did get slowed down quickly, and the noise level dropped significantly. As we climbed one of the pilots told the rest of the cell what had happened and that we had abandoned the low-level route. Number three came out with us to give us assistance, and we leveled off about 8,000 feet. We did a controllability check to make sure all the required things worked as expected. Then number three maneuvered around us at fairly close range for a visual inspection of the damage. We were lucky again—no damage seen except a two by three foot hole in the radome just below the apex. Of course by then we had already seen other indications of the problem. The navigation and bombing radar wasn't working and neither were the infrared or low-light steerable sensors mounted on the airplane's "chin" below the radome.

After explaining our situation to the air traffic center, we were cleared to 24,000 feet for the trip across the Sierras and home to Mather. Passing 12,000 feet we did the usual oxygen check and were cleared off, but in a few seconds my ears let me know something wasn't right. I reconnected my mask and asked the copilot what the cabin altitude was. His response was something like, "S**t! Fifteen three! Everybody back on oxygen!" We had just passed 18,000 feet so we got clearance to maintain that. We spent the next two hours "on the hose" until we crossed the Sierras and let down to home field.

Enroute, our EW used the long-range HF radio to contact our command post about our situation, so they were ready for us on arrival. We let number three land ahead of us and the runway was closed behind us so it could be checked for any debris. As I recall, parking was pretty simple—we just pulled into a designated open part of the ramp and stopped.

After we got shut down and exited the airplane with our personal gear, I was amazed to see the size of our welcoming committee. Maintenance folks were already inspecting the hole in the radome, and by the time I finished my post-flight equipment inspection, they had opened the radome. It wasn't long before one of them found the remains of our bird—a bald eagle! They also found out why our airplane had pressurization problems. The impact had not only pierced the radome and destroyed most of the electronics behind it, but it also punched a significant hole in the pressure bulkhead that sealed the crew compartment from the outside. That hole just acted like another outflow valve in the pressurization system, but it was an uncontrolled one.

We were offered a complementary checkup by the local flight surgeon, but we all passed on that. After maintenance debriefing we did our operations debrief and paperwork (the flight's not done until the weight of the paperwork equals the weight of the airplane), and we got another surprise. The wing flying safety officer stopped by to let us know that the initial assessment of the damage would exceed the then $500,000 threshold for a Class-A accident. Usually Class-A's involved crewmember deaths and/or loss of aircraft, but the Air Force also grouped accidents that were particularly expensive to fix into the severest category with them.

Although our event had been dubbed a Class-A, most of us had been involved with bird strikes before and this one wasn't necessarily the worst we'd seen or heard about. I vividly remembered seeing large holes in wing leading edges, shelled out engine casings, smashed windscreens, and even a bird blood-covered instructor pilot while I was assigned to other B-52 units. But after giving our experience some thought, I knew we had been really lucky to have survived our experience that day.

By comparison, SAC's most recent bomber Class-A accident before this one involved a B-1B (84-0052) from Dyess AFB Texas. On 27 September 1987 it was flying around 560 knots at about 600 feet AGL on another low level route (IR-501 in southeastern Colorado) when a pelican impacted the wing root area. The end result was SAC's first loss of an operational B-1B and three fatalities. I recently discovered that during calendar years 1955 through 2009, there had been 98 B-52 Class-A accidents resulting in 76 destroyed airplanes and 315 fatalities.

The airplane was grounded for many months, but a sharp team from the Oklahoma Air Logistics Center at Tinker AFB came up with an innovative way to keep the repair costs down, and our Class-A accident was down-graded accordingly. The airplane was repaired and went on to fly again until it was retired on 10 July 1989. Our airplane was 57-6468, the very first B-52G delivered to SAC. If you care to see it, drop by the Kenney Gate near SAC's old headquarters at Offutt AFB, Nebraska. You can't miss it—it's the big ugly fella on your left as you approach the gate. It served our nation well, and it brought me home that day. Oh, and yeah, I did pass the check ride.

"He" Was With Us That Morn

Jack Hawley

In retrospect, if a handful of crewdogs were asked what their worst mission takeoff time was, one would probably get a handful of answers. In my case, it was when I walked into the squadron for the morning briefing, and was told, "Go home and get crew rest; the scheduled instructor and/or pilot for tonight is DNIF and we need you to cover for him!"

On the flip side, the early morning takeoff times weren't all that bad since I'm one of those morning folks. Such was the case for a particular mission, with an O-four-dark-thirty takeoff, and a no-show from an EW who was the epitome of a professional. It seems he was running a tad late, put the peddle-to-the-metal, and got stopped for speeding. When those unpaid parking tickets back at college, a couple years prior, showed up on the records check, he was off to the slammer to explain why. The IEW, that was scheduled, filled in just fine.

Let's forego the boredom, of reiterating base ops briefing and "paper war" requirements, prior to proceeding to the aircraft. Many of us have been there, and done that. And likewise, you can all remember how motivated and wide-awake you were, lurking about base ops and the aircraft at 3:00 a.m. in the morning.

The preflight went as planned. About 30 minutes prior to takeoff, we cranked engines, and 20 minutes prior was taxiing to the active runway. The prior-to-takeoff pilot/copilot brief probably caused some to doze off momentarily; however, we had met all requirements and were ready to go.

"Ground, Dozer 24, request clearance for takeoff."

"Roger, Dozer 24, contact tower for takeoff clearance."

"Dozer 24 going tower."

"Tower, Dozer 24, number one for the active, request takeoff clearance."

"Roger, Dozer 24, winds calm, cleared for takeoff, runway 35, contact departure control when airborne."

"Dozer 24, cleared for takeoff."

"Co, I have the aircraft."

"Roger, pilot, you have the aircraft."

We rolled out on runway heading, advanced the power to takeoff setting (which looked good), and started to proceed down the runway. This next sequence covers all of about 10 seconds or less.

At this point, it's appropriate to mention that this pilot is kind of a numbers freak. So here's the situation; its 4:30 a.m. dark as pitch, on a clear, 12,000-foot runway. Ah yes, side runway lights on, so no one needs a depth perception brief at this point, and "Oops!" there's one more light on the right side than there is on the left. Is there a left runway light out?

"Coming up on 70 knots...or...(looks like I get to go to the Commander's office) ... Abort, abort, abort! Throttles idle, airbrakes six, hold the drag chute co, brakes apply, easily. We were approaching Juliet taxiway, perpendicular to runway 35, which was just over half way down the runway at our much reduced speed.

"Dozer 24, this is tower, say reason for abort."

"Tower, this is Dozer 24; there is a large sweeper truck driving down the runway." (Guess where the extra right light came from.) There was silence on the radio for a short time, and then..."Is this Major Hawley?"

"Uh, yes it is."

I quickly found out this particular tower person, was the technical sergeant who had helped me on many occasions as I sorted through myriads of instrument school material, as the sole Base Instrument School Instructor, and his word was sound. His young lad at the console screwed up big time, and at that point, probably saw airman basic in his future.

"HE", the Lord, was watching over us that morn.

"Major Hawley, I guarantee you, I can take care of this one. Is that okay with you?"

It doesn't take a rocket scientist to figure out that there was no one else that was seeing, or hearing, this happen. If my formal report hit the street, heads would roll, big time. No telling how many demotions and firings would have taken place.

"Yeah, that sounds like a good solution; just so that young lad fully realizes what almost happened."

As we pulled off at Juliet taxiway to try again, the sweeper truck casually pulled off the end of the runway, probably totally oblivious, as to what had just happened. We followed up with a safe, successful mission profile. The crew was reminded of the consequences, if this got out.

I never pursued/knew if I was just kept out of the loop or whether we were successful at keeping this "under wraps."

Capt Bushman and 56-0659 with blown tires and skid marks.

Blown Tire On Takeoff

William J. Bushman

This is the story of the flight of B-52D, #56-0659, on 1 Jul 1967 from Andersen AFB, Guam, to South Vietnam and return by aircrew E-70. The crew was home stationed at March AFB, California, and part of the 486th Bomb Squadron that were deployed on Arc Light missions to Southeast Asia.

The crew was made up of the following members:

Aircraft Commander: Capt William J. Bushman (Iowa)
Copilot: 1st Lt Robert J. Rooks (Wisconsin)
Radar Navigator: Maj Albert C. Bartek (Texas)
Navigator: 1st Lt Hollis D. Jones (Texas)
Electronic Warfare Officer: Capt Peter E. Sfameni (Pennsylvania)
Gunner: MSgt Donald F. Pomroy (Colorado)

Our crew was scheduled for a 0200L takeoff as part of a nine-ship mission (three three-ship cells). Pre-mission briefing was held four hours prior to takeoff where each crew position studied his specialized portion of the mission. Altitudes, in-flight refueling details, prospective enemy anti-aircraft, missiles, and fighter possibilities, and targets, etc., were some of the details covered. At takeoff minus two hours a formal briefing with the mission crewmembers was held to ensure all were completely familiar with the night's requirements.

After the formal briefing, aircrews proceeded to their respective aircrafts to perform their walk-around checklists and to pull the safety pins from the bomb load of 108 500-pound bombs. At takeoff minus 40 minutes engines were started and systems were checked to determine their status. If all went well, at 20 minutes prior to launch the aircrafts taxied into their designated formation order at the end of the runway to wait for launch time. This entire flightline procedure was done "radio silent" to prevent a Russian trawler sitting off the end of the runway from monitoring and gathering intelligence about the impending mission.

On this particular mission, we were in the number nine position for takeoff. The CP taxied into position, advanced the throttles to 100 percent, checked all engine instruments within limits, stop watches were started to cross-check acceleration specifications, and the S-1 speed was exactly on pre-flight computed speed. This S-1 speed is extremely important in B-52s because a safe abort for any reason after attaining this speed cannot be made within the confines of the remaining runway. As a consequence, the pilot making the takeoff declares "committed" if the computed S-1 speed is met or exceeded. That means the takeoff must be continued regardless of any mechanical failure or lack of performance.

Right after the CP announced "committed", there was a loud thump. I broadcast over intercom, "Crew we've blown a tire," even though I had never experienced one before. Then the aircraft started pronounced vibrations which got decidedly stronger as each second passed. In addition, the aircraft stopped accelerating and started to wander off the center line of the runway toward the right edge of runway 06R. The RN, who routinely monitors the landing gear on takeoffs through a "periscope," warned the crew that, "The left forward gear is on fire."

As AC, I announced, "I have the aircraft," and fed in full left rudder in an attempt to stay on the runway. Both the AC and CP were on the controls as the aircraft approached the end of the runway and the overrun beyond it. Our ground speed was only 135 knots, nearly 20 knots below the computed S-2, or takeoff speed of 154 knots. Both pilots were pulling back on the control wheel as far as possible while holding full left rudder in an attempt to keep the plane on the prepared surface and "force" it airborne. With a 454,000 pound aircraft this is totally impossible. It flies only when it has adequate airspeed to overcome aerodynamics/gravity.

As we hurtled over the end of the overrun, I warned the crew to "Standby to bailout." Simultaneously the nose of the plane pitched up to a severe angle as happens with swept wing aircraft when they are in the midst of a stall. From pulling back as far as we could with the controls, both pilots pushed the control wheels to their forward limit in an attempt to lower the nose of the plane. I tried to run in elevator trim only to find my control button was not responding. I then called for the CP to "Trim nose down! Trim nose down!" The CP reacted instantaneously and the nose began to lower and the airspeed gradually accelerated. The plane was still shuddering violently from the stall effects. The vibrations were so severe the RN was unable to reach over his head to close the Release Circuit Disconnect circuit breakers to jettison our bomb load and reduce our gross weight. With the airspeed at a critically low point, I had the CP retract the landing gear even though it was still on fire.

We were taking off on the runway that has a 600-foot cliff that overlooks the Pacific Ocean. How much of this 600-foot-drop we used to obtain flying speed none of us on the crew will ever know. (Had we been on runway 24L, we would have certainly crashed into the palm and coconut trees on that over-land heading.) Other crews who had just landed from their mission watched us throughout our takeoff roll and flipping over the cliff. We were gone so long out of their sight that they were expecting to see explosions as we crashed into the sea. They were astonished to see us eventually lumbering our way into the night sky. A later review of B-52 performance data indicated there was no way the aircraft should have stayed airborne at that speed and gross weight.

After we finally accelerated to flaps-up airspeed, the CP extended the gear to blow out any remaining fire in the gear or the gear well. We left the gear down for several minutes before taking a calculated risk to raise them again. As we obtained climb speed, the Command Post made contact with us to check on our condition. They proposed the choices of dumping our bomb load into the ocean and burning our fuel load down to minimums and landing or to continue the mission. The first choice would involve wasting a full bomb load, not hitting a target, and landing in the dark. I chose to continue the mission as planned. When we finally got to cruise altitude and closed into trail formation position we discovered our autopilot was malfunctioning, so we had to hand fly for what would be a 12.5 hour flight.

The remainder of the flight through the in-flight refueling, to and from the target zone, and the return to Guam was uneventful. That is, until we got to the penetration in preparation for landing. Our approach altitude was 43,000 feet. As we lowered the gear to start penetration, the two aft main gears refused to come down and lock in position. At 20,000 feet I broke off the penetration and leveled off to run our emergency checklists. For several minutes we ran the procedures repeatedly with no effect. Finally, after circling in the holding pattern discussing our options the gear majestically came down and locked in position. This was an answer to many frantic prayers because I was preparing to tell the crew we would fly down the runway heading at 5,000 feet and have all but the CP and I eject or bailout. Thank God that decision never had to be made.

After getting all the gear down and safely locked we continued the penetration. All during the flight we discussed reasons for why the aircraft pulled off the right side of the runway if the left front gear had been the one the RN had seen on fire. To figure out what was actually the case with the gear, we leveled off at 10,000 feet and sent the nav back into the gear well area to check out the damage. That is an extremely dangerous task, but all went well for him. He returned to the crew cabin with the news that the right tire was blown and in shreds. It was obvious we would have to land on partial gear.

On getting down to traffic pattern altitude we flew several "low goes" down the runway to allow ground personnel to visually check for damage and to burn our fuel down to minimums for landing. It was somewhat sobering to see the runway lined with fire trucks, ambulances, staff cars and dozens of onlookers. After reaching our fuel minimums, I had the lower deck crewmembers (RN and nav) come to the upper deck for the emergency landing.

With the right forward main gear retracted, the final approach went well, but immediately on touchdown the left aft main gear blew, at 1,000 feet the right aft main gear blew, and then at 4,000 feet the left front main gear blew. We were somehow able to keep the aircraft on the centerline of the runway. Since the right gear was retracted the aircraft heeled over to that side and the nose dropped considerably. As soon as we came to a halt we quickly unstrapped and started to evacuate the aircraft. We were startled to see our lower hatch open only about halfway with the plane heeling over as it was. We had a tough time squeezing through the restricted opening. The hatch had been opened by the gunner who had opened his own hatch, slid down his emergency escape rope, and then ran forward to help the rest of his crew with their egress.

As soon as we exited the plane, the firefighters moved in to control the fire in the blown out tires. The entire wing staff was there to greet and debrief us on the flight and its emergencies. It was very touching to listen to the emotional comments from the crews that had witnessed our near catastrophic takeoff during the night. They all thought we had "bought the farm." It turned out that the brakes on all the gear were welded together from an anti-skid malfunction. With all that, the crew was blessed to survive the conditions we faced on takeoff.

Speaking of blessed, I was certainly fortunate to have such a professional crew to fly with. The RN observed the cause of our initial problem and warned us immediately. The nav exposed himself to a very dangerous situation in the open wheel well in ascertaining the extent of our damage. The EWO played a key role in each of the various emergencies we experienced by immediately initiating the proper Dash-One emergency checklists which allowed the remainder of the crew to continue to fight control of the aircraft. The gunner risked his own safety by repelling down his escape rope to aid the rest of the crew to safely escape the aircraft. And the CP displayed his superior airmanship by helping to control the stricken aircraft and almost pulling into the air with his physical strength and will. Each member knew his specialty well, kept his head in emergencies, and responded in a timely accurate way to them. If not so, we would not have survived this particular mission.

Later on I was chosen as the 1967 Strategic Air Command's nominee for the Koren Kolligian Jr. Trophy. This is an annual award for "outstanding feats of Airmanship by aircrew members who by extraordinary skill, exceptional alertness, ingenuity or proficiency, averted accidents or minimized the seriousness of accidents in terms of injury, loss of life, aircraft damage or property damage." I did not win.

This B-52D, tail number 56-0659, is now on permanent static display in Warrior Park at Davis-Monthan AFB, Arizona, and is dedicated to our crew.

Left to Right - Maj Al Bartek, Capt Bill Bushman, 1st Lt Bob Rooks, 1st Lt Dean Jones, Capt Pete Sfameni MSgt Don Pomroy stand on the skid marks made by 56-0659 on 1July 1967.

B-52D, 56-0659 in Warrior Park at Davis-Monthan AFB, Arizona.

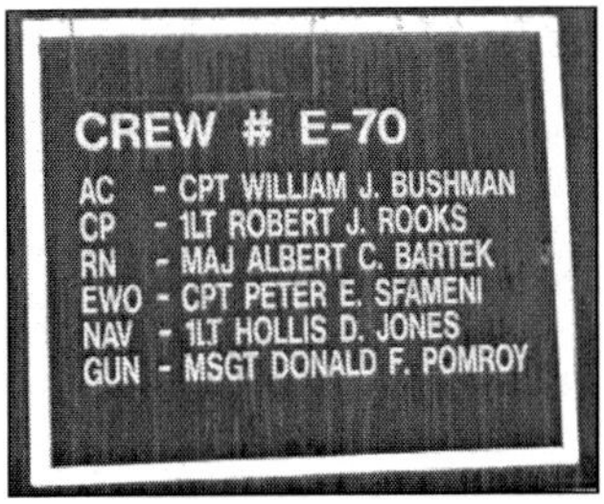

Plaque on side of 56-0659 on display.

First On Scene

Roland R. LaFrance, Sr.

From as far back as I can remember, I had two "loves" in life, firefighting, and aircraft. My love of firefighting came from my father, a professional firefighter for the City of Nashua, New Hampshire Fire Department. I would one day work there myself.

I fell in love with the B-52 the first time I saw one, around the age of six, at an open house held at Pease AFB, NH. It was also the first time I saw, and sat in, an Air Force crash-fire-rescue vehicle.

Growing up, I guess I went through the same thoughts of "What am I going to do with my life?" that most young people go through, but becoming a firefighter was near, if not at, the top of my list. It took a book I read as a teen, a novel about the Korean War, based on the actual experiences of its author to give me direction. The book was (and I still have a copy) *Turn the Tigers Loose* by Walt Lasly, Col, USAF, (Retired). The book is about a group of B-26 Invaders, and the main character is Tom Loring. That last name would come to mean a great deal to me for years to come.

In the book, more than a couple of times, the base crash-fire-rescue units are mentioned responding to aircraft emergencies. That had a great effect on me, and on the rest of my life. Before then I had never thought of having both of my "loves" together. I decided to become a USAF firefighter.

Having enlisted, with a signed contract to become a Fire Protection Specialist, under the delayed enlistment program, I left home for Lackland AFB, Texas in June of 1972, not many days after my high school graduation. After basic training, I was sent to Chanute AFB, Illinois for firefighting school. Although I wished to be sent to Southeast Asia, where I figured my skills would be needed, I was instead sent to Loring AFB, Maine. I was assigned to the 42nd Civil Engineering Squadron Fire Department, 42nd Bomb Wing (H). Strategic Air Command. The flight line was full of B-52Gs and KC-135As. We also had a detachment of HH-3 Jolly Green Giants, and a T-29 trainer aircraft. Loring AFB was named for Maj Charles J. Loring, Jr., a posthumous recipient of the Congressional Medal of Honor in Korea.

It was with amazement that I first saw a B-52 taxiing towards me using its "cross-wind landing system." And I just loved watching the BUFFs take off, come in to land, and do touch-and-goes at Loring. I, and the other firefighters, did a lot of training and ramp patrol, along with responding to emergencies on and off the flight line. But I loved working in and around the aircraft.

In July, 1973, an opportunity came up for a six-month TDY to Andersen AFB, Guam, with Operation Bullet Shot. Someone else on my shift at the fire department was scheduled to go, but I volunteered to take his place. After flying commercial air part of the way, it was on a C-141 that I traveled from Barksdale AFB, Louisiana, to Travis AFB, California, to Hickam AFB, Hawaii, to Andersen AFB, Guam. I was greeted by one of the daily tropical downpours of the summer.

Having arrived at Andersen about a month before the bombing halt of August 15, work for me and the other firefighters was hectic. We had many calls, plus runway stand-by, where we would sit towards the center of the runways in our crash trucks, and watch the multitude of B-52s take off and return every day and night. We also had many tactical aircraft, such as F-4s and F-105s transiting through Andersen on the way back to "The World" from Southeast Asia. As firefighters, we worked a schedule of 24 hours on and 24 hours off. During most nights on duty there were multiple emergencies plus the usual patrol duties, so our sleep was often interrupted.

I remember the days surrounding August 15, and the last bombing mission going out and returning. We still had approximately 150 BUFFs, D and G-models, on the airfield, most of them TDY from stateside SAC bases. Flying continued at a pretty good pace, but our business slowed down some, but not all the way. Our training didn't stop, either. We had pit fires in the fire training area, where JP-4 would be splashed around the old B-52F that was nicknamed, "The Gray Ghost", and we would go in, put out the fires, and simulate rescuing the crews. It was good training, and we were a great bunch of firefighters.

The incident that I will always remember as the most dangerous aircraft incident in my time in the Air Force would occur not long after August 15. I don't remember the exact date, but I will never forget the moment. I was on "ramp patrol", which was driving around our sector of the flight line in a pick-up truck, a P-6, which had one large extinguisher and smaller ones on board. As one of two firefighters, I was riding shotgun. We were operating out of Crash Station #2, on the northwest side of the airfield, and at the north end of "N as in Nancy" taxiway. The station was just an old trailer home made into one large sitting and bunkroom, with a shelter for two crash rigs, P-2s, or sometimes a P-2 and a 1,500 gallon water & foam tanker. Plus we had a P-6 Ramp Patrol Vehicle there. It was while I was riding in this vehicle, this incident occurred.

My driver and I were traveling just near the station when we received an emergency call that a B-52D had landed with "hot brakes" and was sitting in the middle of "Nancy" taxiway. As we turned the corner onto Nancy, we could see a flickering from the area of the aircraft. As we approached it, we could see that the # 1 truck (the left front set of two wheels) was on fire. The hot brakes had caused the brake line to burst, splashing hot brake hydraulic fluid onto the hot wheels and brakes, and it ignited. We were the first on the scene, having already been nearby. My driver stopped at the nose of the BUFF, and I got out, grabbed my aluminized hood, (I already had the rest of my gear on in the vehicle) and put on the hood and gloves as I ran for the crew entry/egress hatch under the nose.

I immediately noticed that the copilot had his escape window open, and his escape rope was hanging from it. The crew inside, six members, did not know whether it was safe for them to egress, due to the proximity of the fire to the hatch. They had one thing going for them. While quite a few ground crew personnel watched from the sidelines of the taxiway, one brave ground crewman was at the hatch waiting on me. He wanted to make sure that it was safe for him to open the hatch. I immediately told him, "Open the hatch, and get them the hell out of there!" Having done so, I stood, as a shield, between the crew hatch and the burning truck. I wasn't thinking of the danger at that moment; I would consider it later. I just thought it was the right thing to do for the benefit of the crew, to shield them from the heat of the fire.

I never saw six men run so fast as that crew did. As they dropped out of the egress hatch, they quickly ran to the side of the taxiway. With the crew safely out, the large crash trucks started arriving on-scene. I gave the hand signals to the first one, to have them hit the fire with the roof turret, which could put out 1,000 gallons of foam or water per minute. The fire was quickly knocked down, and we stretched a hose line to complete the extinguishment.

Everyone was pretty happy. The crew was safe, the fire was out, and the aircraft saved. The 43rd Bomb Wing commander came to check out the scene, and he told us what a great job we had done. For me, the satisfaction of seeing the crew safely egress was all the thanks I needed, but the wing CO's comment was good for morale. It wasn't until after it was all over that I thought of what had happened, and what might have happened. I started thinking that I had a ticking time bomb behind me in the burning wheels. They could have exploded, severely injuring me and the crew, (if not killing us outright) and it may have sent burning shrapnel into the wing and main body fuel tanks. Then, we would have had one large aircraft disaster. But it hadn't happened that way. We had done our jobs as we had trained, and with the help of one brave ground crewman, the crew was saved, and then so was the aircraft.

I spent a lot of training going through the cockpit of the B-52s, mostly the G-models we had at Loring, going either up through the main hatch, or down through the ejection/escape hatches, after having simulated cutting the ejection lines. I had been trained to shut down the aircraft, and the oxygen systems, and knew how to rescue each crew member in each position in the BUFF. I would never have to put that training to use, and I'm glad for that, but I was ready for any emergency.

I only spent four years in SAC, and then one year in the Reserves at Westover AFB, though I still regret not having made a career of it. Loring AFB would have a great influence on the rest of my life. After TSgt Harold Loveless and his family had been transferred to Loring from Tainan AB, Taiwan in 1974, I met, dated, and in February of 1975, married his eldest daughter, Francesanne. We had one child born at Loring, and three in all. One of those is now a 1st Lt in the Air Force, stationed at Luke AFB, Arizona as a Logistic Readiness Officer.

I still think, proudly and with fond memories, of my few years in the Air Force. Some of the greatest memories surround the fire stations in which I worked, and the firefighters with whom I worked. But, also, memories are great of the magnificent B-52, the aircraft, and the crews of which I was trained to protect and save. I still love to see the BUFF fly, and take any opportunity to get new pictures of it for my aircraft photo collection.

To those who still fly it, happy, and safe, landings!! To those who flew it before, during, and after my time as an Air Force firefighter, thank you for all you did, all of your sacrifices, and the great job you did in keeping this country safe and free.

The Vertical Stabilizer of 0060

A Disastrous Chrome Dome

Derek H. Detjen

My personal flight logbook indicates that in July of 1964 our crew participated in one of the first flights to the San Antonio Air Materiel Area (SAAMA) at Kelly AFB in San Antonio, Texas. Kelly repaired and overhauled B-52s for over 30 years. In the early 1960s, they did the major depot-level maintenance workload for SAAMA. Modifications to the B-52s performed at Kelly increased the load capability of each plane and increased the aircraft's range. We left the aircraft there for several major structural modifications, including replacement of the main wing spars and increased metal support straps along the external fuselage. The normal progression then was to take another completed D-Model up to Boeing Wichita, where it would receive the black and green camouflage paint job, among other things. The third leg consisted of flying another newly painted "Vietnam ready" aircraft back to one's home base. Rumors at Turner AFB were that our first newly painted bird was on station for three days before anyone saw it!

I believe that the honeycomb modification to the vertical stabilizers may have been done at some point during this sequence.

It is not communally know, but the need for some of the major modifications made to the B-52 airframe came as a result of an earlier mission flown out of Turner AFB.

During the month of January of 1964, the 824th Bomb Squadron from Turner AFB, Georgia, was sharing its airborne alert commitment with a sister squadron from Seymour-Johnson AFB, North Carolina. This included flying alternating sorties along the eastern Chrome Dome route (encompassing the North Atlantic, Mediterranean Sea perimeter) and returning. Along with the bomb squadron from Robins AFB, Georgia, the three bases made up the 42nd Air Division, who's HQ was based at Seymour-Johnson AFB. On one of the 824th's 24-hour missions, probably the one departing Turner on January 11th, B-52D aircraft 55-0060 encountered unexpectedly strong headwinds on their return trip westward across the Atlantic Ocean. It was determined that fuel considerations precluded them making it back to Turner, and the crew was directed to land at Westover AFB, in Chicopee Falls, Massachusetts.

Since the aircraft was needed back at their home base, and due to crew rest requirements, a KC-135 was quickly dispatched to Westover, to return the fatigued crew and ferrying a made-up replacement crew to Westover to fly 0060 back to home plate. The pilot of the substitute crew was Maj Thomas W. McCormick, the Operations Officer of the 824th Bomb Squadron. The copilot was Capt Parker C. "Speedy" Peedin. Maj Robert Townley, a Standboard radar navigator, and Maj Robert Payne, who worked in the Ops Plans vault, were the navigator team and TSgt Melvin Wooten, a Standboard gunner, flew up front in the EW's position. Although some narratives of the return flight mention the fact that TSgt Wooten simply "bailed out," it's always been my understanding that he in fact rode up front in the vacant EW's ejection seat.

In the early evening of January 12th, the crew received their required weather briefing from the Westover folks prior to their scheduled 1248 departure time, planning to arrive back at Turner AFB at 0315. There has been much conjecture regarding the content and depth of said weather briefing and its adequacy, but no official blame was ever placed on its content. Once airborne on a southwesterly heading, the crew began experiencing increasing turbulence nearing the Cumberland/Salisbury, Maryland, area. The local ground temperature was below freezing and a blinding snowstorm engulfed the surrounding mountains. At 0132 on the morning of the 13th, Cleveland Center got a radio call from "Buzz 14," the call sign of 0060.

Major McCormick informed Center that he was "experiencing light and occasional moderate turbulence and, ah, no problem." A scant two minutes later, Buzz 14 reported "moderate turbulence," and when Cleveland Center queried the aircraft about an altitude change, Major McCormick answered "Ah, roger, we'd like to descent to flight level ah, two nine zero, over." Within two or three minutes, there was definite frustration and concern in McCormick's voice as he said "Passing three zero and we're still in it." The turbulence around the plane was increasing as it slammed into the storm front, and Buzz 14's last intelligible comment was that "We're climbing back up to three three zero." Cleveland Center's positive response was met with garbled noise and no further transmissions, save the controller's comment that the transcript of the reply included "sounding like rushing air." No further transmissions were heard, although Major McCormick did make a frantic "May Day" call and also a "bail out" transmission.

Photo of a different B-52 that also lost the vertical stablizer.

The B-52 Stratofortress Draft Listing webpage tersely lists the accident cause as "excessive turbulence resulted in structural failure. Cumberland, northwestern Maryland." Prior to the aircraft becoming totally uncontrollable, both pilots were experiencing wild gyrations during their attempts to maintain control of the plane. Both McCormick and Peedin specifically mentioned not putting much if any force on the rudder pedals, and yet the primary cause of the accident was the complete loss of the D-Model's high vertical stabilizer, weakened by the excessive winds aloft. Although the last H-Model had been built some two to three years earlier, none of the B-52s had yet been modified with the solid, honeycombed vertical stabilizer interior.

A couple of widely-held rumors at Turner involved the fact that Major Payne had walked through an open barn just before he collapsed and died in the snow and also that TSgt Wooten had perished while hanging in his parachute scant feet from the ground due to the moonless night and lack of visibility. Both of these rumors are totally false, and were quickly dismissed upon reading the excellent account of the accident provided by Lori Ann Haire and the Salisbury, Pennsylvania Historical Web Site. A very brief summary of the recovery efforts indicate that Major McCormick was found by the Maryland Police as he was walking towards a highway that Monday. Captain Peedin survived some 36 hours in the cold and rugged terrain, holing up in his survival equipment before being rescued on Tuesday morning. The snow had continued until Monday evening. Unfortunately, Sergeant Wooten's frozen remains were not recovered until the following Friday, being discovered by Kenneth Holler and his 17-year-old son, Ronald, near their home. Both Major Payne and TSgt Wooten had slid down a ravine into a partially frozen stream during their attempts to reach safety and were ultimately found there. Major Payne's tracks showed that he had travelled almost a half mile away from his chute and then back to it again before finally succumbing to the cold. TSgt Wooten had walked about 140 yards from his chute before he found the partially frozen stream. When the actual aircraft crash site was found, Major Townley's remains were found in the wreckage, still in his seat but not with his harness on.

Air Force and Defense Department weapons experts were quick to caution any civilians in the area to avoid the crash site area until after the two unarmed nuclear weapons were found. They also offered assurances that there had never been an actual weapons grade material accident or explosion attendant to any prior aircraft crash, due to the many safety factors involved in the still unarmed weapons. Since 1950, the Pentagon cited the fact that "there had been some 32 accidents involving nuclear weapons, but no explosions."

Despite these warnings, it was, in fact one Ray Giconi, the owner of the local quarry who got his men together and one of his huge forklifts, a couple of dump trucks and "stopped at a local state boy's camp and took all the mattresses they could find and lined the dump trucks with them." The summary on the historical web site continued that "up at the crash site, they gently scooped up those bombs, one at a time, and gingerly lowered them onto the piles of mattresses in the trucks" before exiting the crash area. A local reporter later asked Mr. Giconi if he'd been scared about carrying those bombs. Giconi's classic response was "I do know that if they'd gone off, instead of being in the quarry business I'd have been in the gravel business."

The townsfolk of Salisbury, Cumberland and the surrounding areas are to be commended for their outstanding and immediate efforts to aid in the crash recovery efforts. Hundreds of volunteer searchers were housed in the Fire Hall, American Legion Hall, local schools and churches. The Lutheran church served over 1,500 hot meals in one evening! Hazel, a local radio person, reported that helicopters "were as thick as flies."

On the ensuing Sunday, a memorial service was held near the crash site, attended by both Major McCormick and Captain Peedin, along with the Peedin family, Scout Troop 140 and Mayor John R. Showalter of Salisbury. At the close of the ceremony, Fr. Robert McCoy, national VFW Chaplain, presented a folded flag to Mrs. Peedin and Scout Troop 140 unveiled a monument, conceived by their Scoutmaster Fred Haselbarth commemorating the accident. Members of the local Fire Department, VFW and American Legion also attended. Mrs. Carol Wooten was unable to attend, due to her small children at home in Albany. The monument reads simply "In memory of TSgt Melvin F. Wooten, of the U.S.A.F. who lost his life on the ill-fated B-52 bomber on Jan. 13, 1964."

Captain Peedin addressed the assembled gathering, his remarks providing a most sobering conclusion to the ceremonies. "I wish each of you had the privilege of knowing Sergeant Wooten as I knew him and worked with him for about three years. The highest tribute I can pay to him is to say that he was an average American who loved his country and served it willingly and well. No task was ever too great for him, nor any task too small." "On behalf of his wife Carol Wooten, and on behalf of myself and other members of the Air Force, our grateful, greatest thanks," Captain Peedin concluded.

Brigadier General Woodrow Swancutt also attended the ceremony and commented that Townley, Wooten, and Payne "gave their lives in defense of this country just as surely as though they had been shot down over enemy targets." As if to punctuate his remarks, a B-52 passed over the gathering at that point, startling the crowd, to which the General remarked "Khruschev understands this type language!"

It was also announced at the time that Major McCormick had just been selected for promotion to lieutenant colonel, and would become part of the 484th Bomb Wing staff, culminating a most distinguished AF career that began as a young pilot involved in the B-29 bombing raids over Japan that finally ended WW II. Also noteworthy is the fact that in March of 1966, Turner AFB and Ellsworth AFB, North Dakota participated in the first-ever six-month deployment to Guam as part of the expanding Vietnam War effort. Captain Parker Peedin led his crew to a successful completion of their six-month combat tour and all of us were proud to have served with him.

Maj Thomas W. McCormick was found by the Maryland Police.

Gear Head

Vince Osborne

It was a routine training mission in a B-52D out of Carswell AFB, near Fort Worth, Texas, on a beautiful day. I was part of an instructor crew, so there were three extra guys (at that time no women were on combat aircraft) on the airplane - an instructor pilot, an instructor navigator, and me, the instructor electronic warfare officer. There may have been an additional copilot because many crews had an additional copilot assigned so they could get more flying time to speed up their upgrade to pilot. I just recall that there was a crowd in the good old BUFF.

As we returned to Carswell all was going routinely until shortly after the landing gear was lowered. Now this airplane has a lot of wheels. There are wheels near the tips of the wings that are essential when the plane is fully loaded and ready for takeoff because the wings are bent down by the weight of fuel, and bombs if it is a combat mission (of course, this wasn't). Then there are the four main gears sometimes called trucks. They are configured unlike any other aircraft in the world and they have some capabilities unlike other aircraft. Using the crosswind crab feature the pilot can aim the gear in a different direction than the centerline of the aircraft so that a landing can be made in a severe crosswind by aiming the nose of the aircraft into the wind with the gear remaining lined up with the centerline of the runway. Each of these four main gears is composed of two large side-by-side wheels. I estimate they reached up to my stomach or chest so they are about 40 to 45 inches in diameter at the outside edges.

In the cockpit there is a light for every gear. Six lights tell the pilot and copilot whether the gear is down and locked when the down-position of the gear lever is selected. In this case the left front main gear did not indicate it was locked. That was an issue. The radio talk with the SAC command post and with the Air Route Traffic Control Center (ARTCC or Fort Worth Center) increased. Interphone chatter increased also. The bombardier has a visual device that he can use to look at some parts of the airplane and he thought the left front gear appeared down, but had no way to ascertain whether it was actually locked. I was somewhat surprised that there was not a definite decision as to what to do when this gear appears down but doesn't indicate locked. It seemed to me it should be in the "Dash-One" emergency procedures section. The Dash-One is the document by which aircraft operations are run. It's an outstanding model for how one should organize critical information and keep it current. And every crewmember was required to keep his document current - every single page of both volumes of it.

It seemed to me we flew around for an hour (maybe it wasn't that long) agonizing about what to do. Finally I keyed my microphone and announced to the crew that I would volunteer to take one of the down locks or pins out to the gear and install it (assuming the gear was actually locked) so that no one would have to worry any more about it. The down locks were carried inside the airplane and were part of the pre-takeoff checklist. The crew chief would remove the down locks from all the gears and hand them to the navigator who would count them and report the number when that step in the checklist was reached. The down locks are designed for the purpose of keeping the gears in the extended position no matter what mistake might be made as a result of human error while the airplane is being worked on between missions.

I donned one of the parachutes, and the navigator gave me one of the down lock pins. I'd never held one in my hand before. But I thought, "How hard could it be? Just find the hole it fits in and put it in." There is a small door behind the instructor navigator's position that leads to the bomb bay, and in case of an unusual situation one could traverse a catwalk through the bomb bay, the 47-section and get back to the gunner's position. And the gunner could make his way forward to the front crew compartment if his pressurization failed in his compartment over a hundred feet away from us. I exited the door and entered the noisy area, kind of a deck with room to crawl (a helmet was a must for several reasons) and I headed toward the left main on my right (I'm facing backwards). When I got to the gigantic, gaping hole where the gear sticks out of the fuselage I was surprised at how much larger it looks from inside the bird than it does on the ground looking up into it. I was also surprised at how many joints and holes there were in the mechanism that not only raises and lowers the gear but allows it to cross wind crab. I had not a clue as to where to put that damned pin. This is where pride, embarrassment, survival, heroism and learning suddenly merged. I thought that if I studied the gear a while and looked at that pin enough times I'd figure out where to put it. Then I looked down and about 1,500 feet below me was Lake Benbrook which is on the approach to the north bound runway at Carswell AFB.

I tried to determine if I could open my parachute and have it fully deploy in the distance between me and terra firma. I tried to determine if I did fall whether I would hit my head then whether the helmet would keep me from becoming unconscious thus making the previous calculation about a fully deployed parachute moot. I thought, "What if the command post tells the pilot to recycle the gear about the time I straddle this hole to put the pin in (somewhere) and the pilot forgets for a couple of seconds that I'm back here." That would have been ugly in many ways including what happens to my body when they extend the gear next time. So, it was time for Plan-B. I honestly studied the gear as best I could to see if I really thought it was down and locked. It did appear to be sticking as straight down as it could be. I went back into the crew compartment and gave the navigator the pin, got on the interphone and reported I could not get the gear pinned, but that it appeared safe to me. The pilot reported that to the command post. With the gear not specifically known to be safe we didn't do the normal numerous touch and goes for pilot proficiency. We just went around again and landed. All was well. The gear didn't collapse.

I wish I could tell you that I got into the wheel well of both front trucks and studied carefully where those pins go so I could save the day in any future situation like the one we had just faced. But that wouldn't be true.

A Most Tragic Weekend

Ray Schalk

Like so many contributors to previous books, I have tons of crewdog memories; but none have stayed with me like these two stories, both of which occurred over one horrific weekend.

The time was the second weekend in July of 1967. I was the copilot on Crew E-39, along with Capt Dick Dixon (P), Capt Ken Palmer (RN), Capt Sal Segreto (N), Capt Charlie Goss (EW), and SSgt George Edwards (G). We were with the 736th BS from Columbus AFB, Georgia and TDY to Guam. It was the beginning of my second Arc Light tour and I had only arrived 10 days earlier. My first tour, back in '66, had seen no losses. I was ill-prepared for the two events of that tragic weekend or the circumstances that had me involved in both of them.

The first incident was the July 7th mid-air collision where two birds on a bombing mission were lost along with six crewmembers on board. I was the copilot in Red 3. The story of the incident is also chronicled by Ron Gable who was in Red 1 (*We Were Crewdogs II*) and Toki Endo in Red 2 (*We Were Crewdogs V*). The second event was the crash, the next day, of Brown II at DaNang where five crewmembers were lost.

I'll begin my memories of the mid-air event. They begin at our mission briefing, which began about 11 PM. The mission called for two cells involving six crews. The lead cell, Red cell, consisted of the March AFB Sutter crew in the lead aircraft, the Columbus Westbrook crew in #2, and our Dixon crew as #3 in the cell. Additionally, the airborne commander for the mission was Major General William Crumm, flying in Red 1. It was to be General Crumm's last combat flight before departing for stateside duty. I don't remember if our crew knew that the general would be on the mission, but I do remember my introduction to him that night. He carried himself like "one of the guys", while leaving no doubt that he was a true professional. My other two memories are that, unlike us, he wore a powder blue flight suit, and that he was dozing a little during the briefing. Now that I am 71 plus, I can appreciate that behavior. With the weight of the ongoing responsibilities of his position and the pending move and the 11 o'clock PM hour, he deserved a doze.

The mission's objective was to strike targets northeast of Saigon. After a pretty routine briefing and pre-flight, we took off at about three AM. Again, all went well, refueling and prepping for the bomb run – donning our chap kits, checking in with the MSQ site, closing in to close visual formation, and such. The Close Visual Formation (CVF) flying part was really fun, for the pilot at least, but everyone recognized the risks. The aircraft separation was, as I recall, wing-tip to wing-tip, nose to tail with 15 feet of vertical separation, which was dangerously close, especially for giant aircraft like BUFFs. The B-52 just isn't as responsive as a fighter.

About four and a half hours into the mission, the trouble began. We had closed to the "V" formation and I was flying the aircraft. Dick and I had worked out that pre-refueling and bombing time would be on me, allowing him to rest to execute the more critical phases. We were in an 80 degree turn to the IP south of Saigon out over the South China Sea. The MSQ called with identification problems and directed lead to change positions with number two. Red 1 verbally initiated the change and number two acknowledged with the intention of descending and moving up into the lead position. At that point, Dick said, "I got it", which I acknowledged. We discussed that we had better give them a "wide berth." I then radioed that we were moving out and slightly back to the left in the formation.

Now I need to digress for a moment. You should understand that this procedure was being executed in a banked turn in a tight formation and in a big aircraft. For anyone who hasn't experienced this, holding your place on lead is difficult enough, even without jockeying around in the sky at 32,000 feet. From our upper position in the echelon in that banked turn, it felt like we were "falling into" Red 2 as he started descending and moving forward. All of that equaled "Pucker time"!

An additional consideration which we observed was that Red 2 had been flying "tucked in tight" since forming up with the other aircraft. Wingtip vortex-generated downwash from Red 1 would have number two experiencing control pressures which they likely used trim to correct. As Red 2 moved down, under, and forward, those pressures changed and would have been neutralized, inviting their left wing to start moving up. Just as they passed forward under number one, the cell roll out was initiated. At that point they were sufficiently close to allow the interacting aerodynamic forces between the two aircraft create an unavoidable collision. SAC Accident Prevention Bulletin 127-1 (Oct 1967) confirms these conclusions.

Psychologists say that "significant emotional experiences" stay with you forever. I can still replay it in my mind. It goes like this.

We were still in the turn – aft and out a bit – when lead started to rollout. "Oh my God!" (Remember how close we all are). I jumped on the yoke with Dick, trying to pull away from the inevitable. We were full left at the stops, but the old bird just responded "oh so-ooo" slowly. It seemed like an eternity until it took effect.

As we rolled left, we watched the two planes fly together. Lead was "rolling out" right onto number two. As they made contact, lead's wing separated from the body as it struck number two. The impact occurred right in the 47-section, severing the tail. I can still see the gunner's white helmet as the tail cart-wheeled away.

Red 1 fell away out of sight. Tailless, Red 2 continued forward while starting into a slight descent. I remember the sight reminding me of a flight of a paper airplane. Just forward of the 47-section, a small line of flames erupted and shot up the center of the aircraft and then out both wings. It looked as if someone had lit a match to a trail of gas.

Then, WHOOM! The entire aircraft exploded in a ball of fire.

Remember, we had recently been refueled by our tankers and we still had quite a load of fuel onboard. James Bond movies were popular then and used huge gasoline explosions for effect. It was just like that! We flew through a huge fireball with debris going everywhere. A loose engine was heading right for our cockpit, seemingly in slow motion. I knew I was dead. Miraculously, it fell away, slipping just under our nose.

Then, just that quick – measured in seconds – the episode was over. I remember the bright sunlight and the eerie quietness – as if nothing had happened.

I have the benefit of adding to this story the additional perspective presented by Red 1's copilot, Wilcox Creeden, as he related his bailout experience in *SAC Combat Crew* (Nov 1967). Here is a quote from that story. "I remember feeling the rush of air on my face and started to believe I had it made. My hopes dimmed when I saw I was approaching a huge ball of flame covering the exploding airplane. As I came nearer to this inferno, I could actually feel the hair on my head burning; I realized I had lost my helmet but it didn't seem to matter now. Luckily I separated from my seat and fell short of the flames".

For us, back in Red 3, the adrenalin kept us focused. My thoughts went to my fellow crewmembers. I realized I had no communication and assumed the rest of my crew didn't either. The last words that I remember hearing were, "Mid-Air! Mid-Air! Mid-Air!" I remember saying to Dick that we were flyable and that I needed to tell the rest of our crew not to bailout. To which, he said, "Okay", and gave me a thumb-back signal to go to the rear. When I returned and strapped in, I found the reason for my communication loss. Between the maneuvering and the fireball concussion, my mike chord must have been jerked loose.

The crew continued to check our air worthiness as we endeavored to finish our mission. Charlie, our EW, worked the radios – especially the HF handling crash information transmissions. The rest of us continued checklists to get the bombs to the target, which we accomplished. Then, as we climbed out, things slowed down and the impact of what we had experienced began to set in. On the way home, no one said anything. We just sat there, reflecting on what had happened, fighting off feelings of horror, helplessness, disbelief, and sorrow. The closest thing to it, for me, is to watch a Youtube replay of the B-52 Fairchild crash.

It was the last combat Close Visual Formation flights for the BUFF.

An additional note is that we did get to visit George Westbrook shortly after his return to Guam. We all talked, but he was still somewhat sedated. His head was, as reported in *We Were Crewdogs V*, swollen like a pumpkin, all black and blue and his eyes were just little slits. He shared with us details of his rescue – struggling to get into the basket while the sharks were closing in on him – his story is something I remember. The rescue pilot dropped his chopper down to the water level pushing George under, thus driving the sharks away and allowing his rescuer to jump in and help. We should give credit to those rescue folks! Credit is due to George's heroism as well!

The second event I want to relate is the tragic Brown 2 crash on July 8th, 1967 at DaNang. This event has not been chronicled in *WWCDs*; and while my story is not first-hand, it is pretty accurate. Pieces have come together to give me some credibility as the storyteller.

First, Maj Swede Brown, the pilot, and his crew were billeted in a trailer directly across from ours, while at Andersen. We knew them pretty well. After the midair, we were grounded during the information-gathering period, and as such, were subsequently assigned duties related to Swede's crash, including handling personal belongings and such. So we were involved. Secondly, our gunner was rotating back to the states leaving us short a gunner. The DaNang crash survivor, TSgt Al "Jugs" Whatley, became our gunner for the remaining five months of his TDY. Finally, a friend of mine, J.T. Chapin, has been working on recounting this crash for a long time. He has dug up many of the details and has been a valuable resource.

The accident unfolds with Crew E-08 from the 736th BS, Columbus AFB, Mississippi, departing on their Arc Light mission from Guam with targets south of the DMZ. The crew commander was Maj Swede Brown (P), flying with Capt Jim Davis (CP), Capt Bill Pritchard (RN), Capt Tony Johnson (N), Capt Don Reynolds (EW), and TSgt Al Whatley (G). After a successful bomb run, they climbed out and leveled off. At that time, things began to unravel. They lost their forward TR units. This meant that AC power was available, but some DC powered switches were inoperable. HF radios were lost, leaving only battery powered UHF communications. Most equipment relying on these switches were out with the important impact on fuel transfer.

There was still enough fuel to get back to Andersen, but it couldn't be transferred around to other tanks. The aircraft was left with only about two hours of flying time. Also, the bird would be pretty heavy when they did land. This had to have everyone rattled enough, and then came more trouble. Just minutes later, they lost #1 and #2 engines. Attempts to restart them failed. Other advice was forwarded to the crew but to no avail. Struggling to keep up, the cell tried descending. It became obvious that there was but one choice for the crew –leave the cell and get the struggling aircraft on the ground.

The decision was made to turn around and go back to Vietnam to land. They picked the base at DaNang which had a 10,000 foot runway. All seemed doable and Swede started his approach. Al Whatley told his story many times when I was present and he always mentioned that they had discussed abandoning the aircraft and "punching out." Al said Swede wanted them (the crew) to do that and that he would try to save the bird by himself. Al also said that they unanimously said "No" and agreed to trust their pilot and "ride her in" with him.

Then the next crisis happened – the flaps wouldn't extend. A lot of people remain unclear on this situation. Between SAC Headquarters, Boeing, and on down, wouldn't the flap extension issue and preparation for a no-flap landing have been considered early on? In any case, it seems to have come upon them as a late surprise and Brown 2 initiated a go-around. They kept the gear down and returned to attempt the no-flap landing.

I'm sure Swede tried to "plant it" as short as possible. With a "no-flap" he would have needed every inch of available runway. Some reports say that the bird touched down, then took a big bounce. Al never mentioned that to any of us, although he did say that he "popped the drag chute." J.T. Chapin reports that eye witnesses had stories that varied somewhat. Exactly what happened on the landing that day remains unclear still. The devastating results are not. The aircraft ran off the end of the runway, sheared the front trucks and hit the ground sliding - disintegrating as it went – right into a mine field.

Fortunately for Al, the tail sheared off leaving Al trapped facing the runway. The aircraft burst into flames, followed by explosions resulting in the loss of the five forward crew members. Again, Al told us that he remembers sitting there, resigned to die, as the flames lit up the sky behind him and consumed everything else.

Now for Al's miracle story.

He said he could see firemen in suits, red lights, and vehicles. He wondered why they weren't coming his way. He wasn't aware of the mine field issue, of course. Al told us it was dark and that he heard a noise. He looked up and saw a guy trying to "chop him out" (I think I remember that the tech order says this can't be done). But there he was, chopping away with an ax. Al wondered how he even got up there. He got Al out as other help arrived.

I remember distinctly that Al said he asked the guy how he could have risked going into the mine field. He got the simple answer that it was obvious to this hero to simply run down, following the skid marks of the severed tail section. The collaborative research collected by J.T. Chapin has other variations of the details of Al's rescue and they may be more accurate. However, these are my recollections. Chapin also reports that Al's first concern on egress was, "How's my crew? How's my crew?" I would confirm this. He always said that his first worry was about those other guys. Familiar theme, huh?

I would add that Al insisted that after the crash it was his choice to stay with the squadron, on Guam. There were concerns, in those times, about morale back at home when something like that happened. I really can't say. After flying with Al as a fellow crew member, I surmise that there was no question for him, one way or the other. If you knew Al, you would completely agree.

One other important sidebar needs to be mentioned here. A tribute should be made to honor the strong women who supported the crewdogs and their mission. "Not enough has been said!" We know what they did on the home front with children, et al. Others don't! I would say that they experienced the same anxiety and fear that we did. They did this day to day, and in some cases, to the extreme – like when a disaster struck. Some readers may not know this. The AF tried to prepare them for the shock and grief that might come. They were briefed, making a list of friends to be there for support, and such; but it wasn't enough. Consider a caravan of blue AF sedans entering the base housing area, with the lead displaying the CO's flag. The "notification party" - also carrying the Chaplin and designated friends - would slowly proceed down the street. The wives in each house prayed it wouldn't stop in front of their home. As it passed, after initial relief, they would then wonder who would meet this awful fate. If it stopped at a house, what would one do? One wife was unloading groceries after her commissary trip. They stopped, to which she threw her groceries in the air, ran into the house, and slammed and locked the door. "Gasp!" Whether you were this wife or any other, you were affected. This kind of worry always hung over their heads. They were a group of understanding women who were truly brave and tough. We salute you ladies!

I will conclude by saying that my crewdog days provided positive growth experiences in my life - more than any other non-religious, non-family experience. My contribution's pale compared to what others did and gave. I appreciate this platform on which to write and share these events. Thanks to all of you fellow crewdogs and all those who supported us as well.

Chapter Four–Southeast Asia

Southeast Asia [south-eest] [ey-zhuh] - *noun* - The countries and land area of Brunei, Burma, Cambodia, Indonesia, Laos, Malaysia, the Philippines, Singapore, Thailand, and Vietnam.

John Bair and the Nav's

Arc Light and Crew Integrity: Staying Together

John Bair

My one and only Arc Light tour to Southeast Asia, during the summer of 1971, was a completely unique experience for a naïve and recently married, 23 year old 1st lieutenant. It was the first time I had been outside the continental United States and we had received some training about the people and culture of Thailand, where we would be stationed. We were instructed that they were very friendly and that we were their guests; however, we were advised that there could be infiltrators from neighboring communist countries and that whenever we went off base from U-Tapao we should stay together as a crew if possible. We were definitely not go anywhere alone.

In general this was a fine policy and helped to bond the haphazardly thrown together crewdogs into a smooth running "pack". My own crew, Kincheloe AFB E-07, was commanded by Capt Ruben Siebert, who had previous Arc Light experience. He led us on many off-base excursions to tour Buddhist Temples, local seacoast resorts like Pattya, and numerous 'booty" trips by baht bus to Sattahip to procure everything from teakwood elephants to jewelry and brassware.

I have to admit it made you did feel somewhat safer when you traveled together as a crew. There is always culture shock in a foreign country where you don't speak the language, and being with five others who speak your own certainly helps. Yet the Thai people, in general, were very friendly and seemed to appreciate us being there.

I remember on one excursion to Sattahip we stopped at a tailor shop. I thought I'd have a three piece suit made and was immediately impressed with their "can do" attitude. These guys were all over me with tape measures while scribbling down measurements on a note pad. I picked out the material I wanted from a batch of samples and voilà, in a week I went back to pick up my new $35 suit. On the inside pocket was an embroidered label: James S. Lee & Company - Champion Clothes to the Champions of Freedom - Custom Tailored in Thailand. I still have the suit after 40 years, and it still fits!

New Suit

Most of the crew also had "elephant hide" boots made. In that particular shop all you had to do was stand on a piece of cardboard and they would trace your foot. You picked a style and type of leather, and a few days later returned to pick up your new $8 shoes or $12 boots!

We were told before we went over there that the Thai people loved to barter, and in fact expected you to barter and might even be insulted if you didn't! Well I must admit that wasn't my forte. It seemed every time I tried, I'd get this: "Ooooh, number 10 deal GI for number one item." The best wheelin' and dealin' I was ever able to accomplish was for an opal bracelet for my wife. They were asking $60 and finally after half an hour agreed to $55. I didn't think too much of that bartering job until years later, back in the States, when it was appraised for $800.

Quite often crews would coordinate with other crews who were going on R&R to see where they were headed. I had a friend on a crew headed on a buying trip to Okinawa pick up a Yamaha classical guitar for me for $35. That was not a bad price for something that would cost $210 back home.

We usually ate meals together at the O' club on base. There they would serve iced tea with slices of ma naou (lime) rather than lemon - something new to me. I once complimented the Thai waitress serving us there on her dress, mentioning that I thought my wife would like that style. One week later she brought me back a newly tailored one. I guess that explained why she inquired about my wife's height!

Often we would stay after eating at the O' club and listen to a local Thai band performing all the popular US rock hits, even singing in English; however their "R" and "L" reversal problem made for a rather humorous version of "Proud Mary" with "Lollin' on the liver..." being the refrain.

Staying together as a crew was fine for shopping and sightseeing but wasn't always my personal preference for where to go or what to do. But I suppose for crew integrity the individual often does have to make a sacrifice if you're blending the values and tastes of six different people. So there I was, a young, naive, religiously brought up newlywed! What were they thinking when they dragged me into that strip bar on the way back from Sattahip? "Come on EW, got to stay together!" I didn't even drink, so I "nursed" a $5 Coke, with my back to the stage until the band got quiet and the whooping and hollering suddenly died off. Someone tapped me on the shoulder and I turned around to find a bare-breasted stripper who proceeded to flop her boobs in my face! I was so shocked I jumped up and ran, red faced out the door. I can still see my copilot, Wayne Merritt, practically falling over backwards laughing. Looking back I can imagine how ridiculously funny it would have been. If they only had video cameras back then, it would have definitely won America's Funniest. Regardless of crew integrity or "Comm Sym Pinko" infiltrators outside, I was out of there. I was definitely a married man and had absolutely no outside interests. I would certainly want my wife back in the States to feel the same. Red faced or not I am still proud of that strategic retreat!

Places on base did provide some opportunities for solo activities during my off-duty time. There was a lapidary shop where local Thai craftsmen would teach people how to cut and polish gemstones. I learned how to make opals: then I'd take them to Sattahip and have them mounted in 18 carat gold settings. Now that was an activity of which my wife genuinely approved. When I got back to the States I made my own little lapidary shop and would order rough opal from Australia to continue the new found hobby.

I guess the only time we split up as a crew was during our six days of R&R. The AC and Capt Al Shultz, my radar nav, had their wives fly to Thailand and they spent their off time at Pattya on the coast. I hopped a homeward bound KC-135 to meet my wife in Hawaii for three days: Wow, a second honeymoon! (And I behaved myself as can be seen from the following photograph!)

As far as I know, the copilot, nav (Capt Wayne Witherspoon), and gunner (Msgt Art Pollard) stayed behind. Maybe they deserve the Arc Light Crew Integrity Award!

Panel 2E, Row 12

Dave Wellman

There isn't a bomber crew member around who hasn't heard of the famous August 1943 B-24 bombing mission of WWII of the oil refineries near Ploesti, Romania. The mission, flown at low altitude, turned into one with big losses for the AAF. Some of the losses were attributable, in part, to navigation and timing errors introduced by the unfamiliarity of low altitude formation flying.

Fast forward to 18 June 1965 when the first mission for SAC's B-52's in the Vietnam War was flown. Incidentally, the recently deceased Nguyen Cao Ky, flamboyant Vietnamese Air Force General, was named Prime Minister of South Viet Nam on that same day.

Units from the 320th Bomb Wing, Mather AFB, CA and the 7th Bomb Wing from Carswell AFB, TX were deployed in early February to Andersen AFB, Guam. The entire complement of 30 B-52Fs (plus two spares) were loaded and poised to make a major strike. The target was the Phuc Yen Airfield northwest of Hanoi and was scheduled to be struck at low altitude.

In reality, this was not to be. Instead, crews busied themselves with occasional training flights, formation flying practice, RBS runs against the Nike site on Okinawa, and live bomb drops on Nafton Rock near Rota Island, just north of Guam. Leisure time was spent enjoying swimming, tennis, volleyball, and a frog jumping contest. Yes, a frog jumping contest! Much unlike the California Calaveras County "Jumping Frogs," these were Guam jungle toads! Rescued from the hot grass areas outside the crew quarters, watered and kept cool to keep from dehydrating, the toads provided a wagering outlet for crewmembers as they were contested in the Base Ops firehouse. Crews were far from thinking anything warlike was ever going to happen.

Then it did. The same B-52's were again readied for a strike, not at low altitude as planned, but high altitude. And not Phuc Yen, but Ben Cat, a jungle Viet Cong staging area in the Michelin rubber plantation some 40 miles northwest of Saigon. The target was a one by two nautical mile "box". The French, still having a large influence in what was French Indo-China, complained throughout the war about damage done by bombers to their plantations and demanded tribute be paid. How would you like to have been in LBJ's office when he told the French ambassador what he could do with his rubber trees?

The tasking order called for 30 B-52's in two wave groups, consisting of five three-aircraft cells each. 24 aircraft were loaded with 51 750-pound M-117 bombs and six aircraft were loaded with 24 M-117's externally and 27 1000-pound bombs M-59s in the bomb bay. The plan called for an air refueling with 80,000 pounds offload; trail formation; and bombing altitudes between 18,000 and 22,000 feet. Anxiety ran rampart as crews readied themselves for takeoff that would put them over the target near 0300 Vietnam time. The specialty briefings included very specific instructions for navigation and refueling. The mission was to be flown in radio silence.

Using 30 aircraft in two 15-aircraft waves was not something covered by Tac Doctrine! B-52 crews were used to flying single aircraft Cold War training missions. This mass gaggle was to present a challenge to stay in place in the stream by station keeping using the RN's radar within each of the cells following the rendezvous beacon of its wave lead.

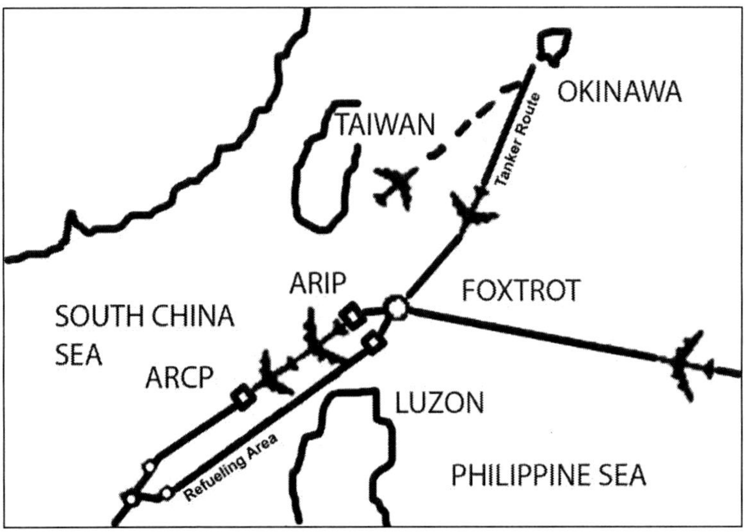

Figure 1- Control Point FOXTROT on 18 June 1965

The tragedy actually started with the pre-takeoff briefings held in the fire station on Andersen. One never-to-be-broken cardinal rule was that once they hit "Foxtrot", the fan point for all five refueling tracks, (See Figure 1), an airspeed of 440KTAS was to be held until the ARIP, followed by point-parallel rendezvous procedures. This was a point hammered home by then 3AD Commander, General Ohlke.

Gen Ohlke, was seldom if ever seen around the Arc Light Task Force environs, but was best known for his personal and hand polished B-47 which sat gleaming on the side of the flight line occupied by the B-52B's from Biggs AFB on nuclear alert. Ohlke insisted on not only making "Foxtrot" on time, but that all waypoints would be on time! On that night the weather had turned foul and a typhoon was situated right in the middle of the refueling area west of the Philippines. Even the Coast Guard had pulled some of its boats out of the area. For this mission, a two-engine amphibian rescue HU-16 Albatross "Duck Butt", was placed on alert at Clark AFB in the Philippines.

The B-52's pressed on, experiencing 50K+ tailwinds aggravated by the airflow from the storm instead of the predicted light and variable winds. They were going to get to "Foxtrot" very early and the tankers would not be there. They slowed as much as possible, but arrived at "Foxtrot" as much as 23 minutes early - difficult but workable.

One cell leader lost its radar inbound to "Foxtrot." In that situation the procedure was to do a 30 degree right, 60 degree left, and 30 degree right maneuver and fall in behind the 3rd aircraft. That would make the 2nd aircraft in the cell assume the lead and navigation responsibility. This reformed cell proceeded to "Foxtrot" and headed to the ARCP for the right refueling track.

It was on that leg that the new leader of the cell decided to arrive at the ARCP "on time" apparently encouraged to do so in the briefing. The tankers from Kadena had not overtaken the bombers at that point. A 360 degree maneuver was called for and halfway through the turn, the turning cell came on a collision course with the cell that was due at the ARCP for the adjacent second track 15 minutes later. The number two aircraft in the turn collided with an oncoming aircraft. One should realize that the weather was not ideal and spotting another aircraft coming directly toward you at night was problematical if not impossible. To the best of the surviving crewmembers' recollection, the turning aircraft clipped the vertical fin of the aircraft in the following adjacent and incoming cell. Seven crewmembers died in the collision, including both gunners. Five crewmembers ejected successfully, but one of those, pilot Joe C. Robertson, died from loss of blood about the time he was rescued by a second HU-16 which replaced one that was damaged when attempting to set down in the high waves.

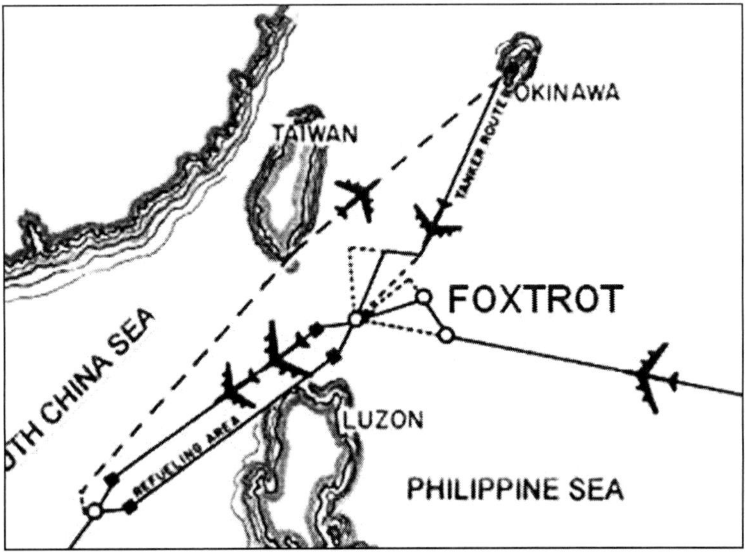

Figure 2 - Control Point FOXTROT with Timing Triangle for Bombers and Tankers

A total of 27 B-52's completed the bombing mission with one having to abort into Okinawa due to a hydraulic problem. One B-52 had to land at Clark AFB on the return flight to Guam. Flights were stopped for a couple of weeks while the cause was determined. The main "fix" was the addition of the navigation timing triangles for both bombers and tankers leading into "Foxtrot", a practice followed throughout the remainder of the war. (See Fig. 2) Restrictions on the use of radio transmissions were also revisited. The bombers in the Ploesti raid conflicted with each other in the target area having been confounded by turning point identification problems and unfamiliarity with directed tactics. The bombers in the Ben Cat raid had a different set of similar difficulties.

About a month after the accident, a memorial service was held in the packed and overflowing Mather AFB Base chapel for the lost eight crew members. At the same time the service was being conducted, my crew was over the China Sea at the position where the collision occurred.

There, in the dead of the night, at 10,000 feet, we "laid" a flower wreath which had been taped to the right bomb bay door.

May God Bless the memory of our lost fellow crewmembers whose names are found on the Vietnam Memorial, Panel 2E, in the vicinity of Row 12.

James M. Gehrig
Tyrrel G. Lowery
William E. Neville
Joe C. Robertson
James E. Marshall
Robert L. Armond
Frank P. Watson
Harold J. Roberts, Jr.

This story is dedicated to the memory of my close service and business friend, Lt Col Roland Larson (deceased), whose dedication to his country, his family, and to God remains an inspiration to me and many who had the privilege to know and serve with him. His recollection and research contributed to this vignette. For an excellent and personal account of this fateful mission, including the heroics of the crews of the KC-135 tankers which couldn't offload, read "*Arc Light One*" by Don Harten, the surviving copilot.

The Wall contains the names of the other B-52 crews as well.

The author briefs Gen Lew Allen, AFSC commander, on an earlier flight.

Did You Hear That?

Clif Banner

Preface: It's been a long time since my early B-52 flying days. For my first combat missions, I was fortunate to have been assigned to a great Westover crew with lots of Southeast Asia experience. All told, my 13 years of continuous flying were very good years; having entered the Air Force as draftee with a lottery number of 10 - flying for the Air Force was a faraway dream - once in the cockpit, I never looked back - what a ride!

Today, I'd love to be able to get back up in the BUFF once more - do some refueling, touch-and-go's, low-level flying, and, of course, the bombing mission; I'll not miss the long high-altitude navigation legs, however. Besides being the only means to transit to home station after flying half-way across the country for an obscure oil burner low-level route, high altitude nav legs were very boring for the pilots— much like the same experience of passengers in the cabin of a commercial airliner. But, pilots performed an important role during the SAC high-altitude navigation training legs, by keeping the heading and speed exactly as the nav team needed it. This segment of the flight allowed the pilots to finally get a chance to eat our aging, stale box lunches prepared by the flight kitchen the night before; and, to prepare some of the post-flight paperwork, finely tune our onboard systems (fuel, electrics, engines, etc.), while continuously scanning for air traffic and thunderstorms.

Air Route Traffic Control Center (FAA) usually let SAC bombers drift as much as they needed to provide critical navigation training for the wartime mission. Since the majority of my navigation training legs occurred while I was stationed at Griffiss AFB in Rome, New York, our homeward bound tracks generally took us close to the Canadian-US border. We'd have flown our low level bombing mission in the Bismark (North Dakota) or one of the other western ranges, and ARTCC was especially sensitive when we approached the national border. Our deviations from the filed route during some nav leg, when the navigator was blind to our actual location, were, at times, so great that we'd re-file a few points with FAA when the Radar Nav quietly requested it over private interphone. During that navigation leg training, the navigator wouldn't hear any clues concerning his real (vs. celestial and dead-reckoning plotted) location.

This story is from my B-52D days, not long after I finished Castle Combat Crew Training School (CCTS) in the B-52F model, completed D-Difference (about three flights in a B-52D) at Carswell AFB, Texas, and deployed to Guam. The crew would spend a few days on the ground at Andersen AFB, Guam, before commencing combat operations. My crew decided we'd be better off at U-Tapao Royal Thai Air Base (UT or U-Tapao RTAB). Guam was good if you preferred the Western cultural experience (Navy BX, some imported rug merchants, and the like), but we preferred the in-country Asian feel of U-Tapao, its Thai culture, and, frankly, the shorter combat missions. Why fly 13 hours when you could fly six? We were one of several crews with more time spent in Thailand than on the island. Our EW, Ed, was married to a Thai woman who acted as our translator with the Sattahip merchants, took us up to Lop Buri (her home town) by train (definitely, a culture shock experience), and allowed us far wider travel experiences than the island-bound crews.

Flying U-Tapao missions, we had, in a way, relatively-short nav legs coming back to the high fix for U-Tapao upon completion of our bombing missions. And, our mission target areas varied a lot—as much as an hour farther east on some nights than others. I flew, at the most, one or two bombing missions from Guam, and they usually ended with a landing at U-Tapao, where we'd stay for a few months at a time.

Our missions out of U-Tapao Royal Thai Air Base (U-T) normally started around 2000 local. I recall our schedules were "On three, off one" or "On five, off two", and we looked forward to those off days for a chance to hire a driver for an all-day trip to the famous Pattaya Beach on the Gulf of Siam or to Bangkok. As SAC bomber crews, our free time was very limited and controlled; we were told to travel in packs (there were occasional knife attacks on lone GI's), but we could hire the "Baht-Bus," a highly-decorated mini-pickup truck with bench seats in back, or a sedan taxi, which the drivers adorned with lotus blossoms hanging from the rear-view mirrors wafting their scent across our faces. For $20 bucks, we had the driver all day, departing U-T around 0900 and returning well after midnight. We'd normally have four or five of the crew along on each trip—one normally staying behind to work out his weekly bout of dysentery, sleep in, or head to the Club to "re-hydrate" early.

Pattaya Beach and Bangkok hotels were four-star or five-star, with great food, good entertainment, and lots to do. Two bucks would buy you top shelf liquors almost anywhere, but, of course, as trained SAC aircrews, we never overindulged. Germans and Northern Europeans flocked to Pattaya's fine beaches for the sun, relaxation, and nightlife. That created a strange feeling in most of us—there, we Americans were on a rare day off from "the war" and all around us, other countries' citizens were enjoying the blessings of peace and prosperity, seemingly oblivious to the American warriors scattered around Thailand.

Live entertainment at the hotels and restaurants was amazing. The bands and singers, most often from the Philippines, would be performing the latest American Top 40, singing them extremely well, and closely mimicking the original bands even though they rarely spoke any English. Occasionally (always), there'd be at least one rendition of "Yarrow LIvel" ("*Yellow River*") and "Lorring on the Livel," better known as "*Proud Mary*," if the band spotted GI's in attendance.

Well, back to our flying story….

Mission show time had us all forming up in the briefing room.

"Red flight, your targets tonight are bicycle storage, fuel storage, and bivouac areas along the Ho-Chi-Minh Trail." (Groan! Not again!)

"Com plan is as briefed for the week. Pilots, call out your pre-release alerts on Mike-2 channel and Guard: 'Attention, Attention—our bombing run will occur in four minutes, Bearing 345 degrees, 22 miles on (TACAN) Channel 72.' You're holding in Delta North (timing box), and you'll depart Delta-4 for a time on target (TOT) of 1930 (0230L). Expect any inflight mission change orders NLT takeoff plus 1.5 hours. Takeoff interval is 30 seconds. In event of an emergency during takeoff beyond decision speed, fly out to 4,000' and drop your bomb load, armed, into the designated safe area of the Gulf (of Siam)" But, that's another whole story—I'll just say that with the amount of hours we were accumulating on the D, shelling an engine was not uncommon on takeoff; and when we dropped 108 500-hundred pounders into the Gulf, the fish would magically pop to the surface—a few sweeps of their nets, and the fishermen's night fishing was done.

On the flightline, we were all business; with wingmen two and three departing 30-45 seconds after lead. Dozens of planes would form up in three-ship cells flying a loose echelon formation, then moving to station keeping in trail. We stayed at strictly-briefed intervals for the specific bomb loads and intended release patterns. Our positions in trail were confirmed by the gunner of the lead aircraft so that our nav team could keep their eyes on the scope for the few recognizable radar return features by which to navigate.

The ability to navigate in Southeast Asia using radar offset navigation was quite difficult. Things improved rapidly when our Westover crew was selected to train and fly the Loran-equipped B-52D aircraft, called Pave Buff, and we alternated by flying missions with similarly-equipped Pave Phantoms (F-4 with Loran). At that time, we had no GPS, no INS, generally no celestial, lousy counters, and not much to see on the ground—thousands of square miles of jungle canopy—just a few hills, coastal lines or water/canopy contrasts. On some missions, we employed pencil beam Combat Skyspot radar control from surveyed, fixed ground sites, guiding us to target drop points on some missions where there was nothing else beyond TACAN to know where we were.

On my first tour, new to the SAC bomber crew force and flying right seat, the radar nav (Bombardier), Bill, was all over me on the first few missions to keep on airspeed (plus/minus ½ knot—he wanted it pegged solidly at the assigned indicated airspeed. Pete, our aircraft commander, had me moving the fuel – "Co, you've got 3,000 pounds more in the left outboard than the right." I'm all assholes and elbows—

a new pilot's curse—pray you don't have an electrical alternator (pneumatic) problem at the same time.

As predicted, 40 minutes into tonight's flight—headed toward the timing box—we get our mission changes. Just follow the lead? Not this time, our flights are split and the wave is split—the targets Washington figured out last night had moved—we have two mission priorities now, and we're on our own, now, as lead with Red 3 in tow as number two. Red 1 joined Blue flight to fill the hole left by an aborted takeoff on the priority mission.

Later, once we've completed our bomb drop—Bill, of course, sees numerous secondaries—I guess the bicycle tires were filled with fuel? "Smart-Ass," is Bill's rejoinder. We're on our way back home—RTB. The nightly thunderheads are reaching above 45,000 feet—tall, wet, dangerous, and all around us. Some thunderheads reach up to 60,000, it seems and are rapidly depositing their deluge onto the jungle below, rapidly shrinking and revealing low cumulus cottonballs scattering over the eastern flatlands. The first change of the night sky from jet-black ebony to deep, dark navy blue appears—the bright stars slowly fading as the eastern sky begins to lighten.

By then, it's 0442 local when we first hear it. Just a faint, distant whisper: "Did you hear it?"

Five minutes go by: "Did you hear it?" Was it on Guard, Tactical, or Nav radio? Three minutes pass. Again, "Did you hear that?" At our altitude, it could have come from the east coast of Vietnam, the center of Cambodia (no—we weren't really there, were we?), from some hilltop radar site, or from another in our disbanded wave of U-T bombers. At that time, there are no commercial flights—most of the fighter and tanker activity is completed for the night, and the stress of the mission is mostly over.

No other sounds—just the whoosh of air slipping by at 47,000, kind of quiet and the cabin altitude is well above the standard 7,500 feet. The oxygen is flowing, our ears are slightly deafened by the high altitude flight, and we are just navigating back to U-Tapao high fix—almost time to step down to the mid-30,000's. The dark navy-blue horizon is now light blue—still dark to the west, but as we look over the wings, we feel the rosy glow of the sunrise—red sky in the morning? Good thing we're not on the water.

One more time—just a bit louder this time: "Did you hear that?" Some other airplane finally breaks the silence: "Did we hear what?"

In a God-like, booming, echoing voice—deep and rich: "The CRACK of Dawn!!"

Yuck! Yuck! Yuck! Har! Har! Har!

We probably heard that a hundred times over the theater campaign—and, yet, it's still funny.

Now, U-T Approach Control is on the air, "Red 2 Flight, descend and maintain FL 230." Ed, our EW, is on the Command Post UHF frequency, calling in the night's numbers. We're headed down. As we hit the humid U-Tapao air, our windscreens instantly ice over—windscreen heat is on high, but we're blinded for a few minutes as the cold-soaked airframe begins to drip. The throttles are back, have been idling for the past 22,000 feet of descent, and not much air conditioning coming into the cockpit. We're on instruments for about three to four minutes as we part the clouds and see the early morning puffies dotting the Gulf and U-Tapao surroundings.

Checklists are completed. The nav and radar had been debating which of them—or, perhaps, the copilot—would check the bomb bay for hangers (hanging bombs—partially released). Bill, the radar, takes it tonight as we descend below 12,000 feet. He's grousing about the temperature and the dripping air, yet doesn't forget to highlight tonight's secondaries.

We'll do it all again tomorrow—and the next night. Different call signs, different timing box—maybe, a mission uninterrupted by a mission change (NOT!). But, now, we need to shut her down, debrief, and wander over to our air conditioned crew trailers to sleep off our night's fatigue.

"Did you hear that?" Ha, ha. Wartime humor…it keeps us going.

Port Bow and More

Mike Radke

I consider myself fortunate that I've always had good relationships with my crew members. Apparently it was not that way for all crewmembers. I was the gunner on my second Arc Light tour from Glasgow on a new crew. The pilot, Tom, was a major. The copilot, a captain, was an Academy grad and a very quiet reserved individual. The nav team, Jim and Ron, were both captains and a rowdy pair that was always at odds with the pilot. Gary, the EW, was a captain who minded his own business. The crew relationship always seemed to be strained for some reason. They just didn't get along. Maybe it was because of their first tour together.

On most of my tours, I was always lucky to fly cell lead, but sooner or later luck runs out. On a flight in September of 1967 we did the dreaded bag drag to the unmanned spare. To our chagrin we had no radar scope at all. Maintenance said it was on the way. Number one and number two had launched and a Bomb/Nav MSgt maintenance man was on board working on the scope when we launched. Because of the delay we were vectored across Laos to catch up.

We flew north of the DMZ and had flown across the same SAM site five days in a row. We closed to three miles at IP and because of having no operational radar scope we released our bombs at three miles by a Bonus Deal from number two. Outbound we overflew the same SAM site and they launched five SAMs at us. At the time there had been no missile maneuver procedures devised. From somewhere out there we heard "Don't worry about the piss-ants, the elephants are stampeding," and I got it all on film.

I do not know what happened up front in the cockpit, but there was a lot of swearing and the copilot said he had the controls. There must have been some sort of confrontation, because there was still a lot of tension at debriefing. The Bomb/Nav troop probably would never want to fly again. The OSI confiscated camera and instructed us not to discuss the episode. The next day the *Stars and Stripes* headlined it. Go figure.

Also, the next day I saw my pilot, and he was a mess of bruises but wouldn't discuss it. Gary, my EW, told me I really didn't want to know what happened. The nav team's rage continued until we completed our tour. The pilot avoided the crew on every occasion for the rest of the tour.

I was so thankful that tour was over and we flew home penetrating in a blinding snow storm and landing without incident. Everybody quickly departed to get to their families. Upon our return less than a week later, we were notified that the USS Pueblo had been captured by the North Koreans and we were leaving immediately for Kadena AB, Okinawa, for an operation called Port Bow (1-23-68 to 2-13-68).

Upon arriving at Kadena, we were put on hard alert, with real tight restrictions. Nobody was to leave base. It was rumored that some of the support personnel violated the restriction and were sent home for discipline.

On our first morning's pre-flight inspection we experienced our first SR-71 (Habu) takeoff. At that time it was a plane we didn't know existed. As time went by crew morale declined due to boredom and many suffered from dysentery from the food or water.

Every day the political demonstrations outside the gates grew and we inquired about why our alert bus drivers were participating. They told us they only participated because they were being paid very well to do so. As time went by and as long as we were there, the staff decided we might as well start flying sorties over Vietnam and so a few crews at a time were allowed to go into town. The missions were about eight hours, a lot shorter than Guam. The Flight Surgeon supplied us with what we called "eight-hour corks" for the diarrhea problem, but if the flight took longer – too bad, you had a problem.

Our rotation for town arrived and Gary, my EW, and I did our own thing. He became good friends with one of the bar maids at one of the watering holes. He was single. She claimed to be a school teacher, working in the bar for extra income. It got very serious between the two of them and soon the inevitable happened. At some point Gary lost his flight bag and we went to the OSI Office to file a report of the loss. The receptionist sent us back to an office and he knocked on the door. The female agent opened the door and what a shocker it was to see his girlfriend standing there! It was a good thing he had not divulged any crucial information during their relationship. Shortly thereafter the 322nd rotated to U-Tapao and we fell back into the old routine. We were housed in the Texas trailers at the time. The EW and I had lock troubles on our door and we slept with our 38 pistols. Nobody trusted the Thai Marine Guards at that place and time. I awoke one morning and all I saw was a switch blade knife, and almost shot my EW. He was trying to repair the door.

At the end of May, 1968, Glasgow closed and everybody departed to the four winds. Glasgow and the 322nd still have many good memories for me.

Sawa-dee!

A Few War Stories

David C. Wellman

(The following three stories are from late December 1965 to March 1966 and happened while on the second Arc Light deployment of the 320th Bomb Wing, Mather AFB, California.)

Thanks for the Memory

On 10 April 1987 I wrote the following to Bob Hope:

Mr. Bob Hope

Palm Springs, California

Dear Mr. Hope,

On a recent vacation trip to Palm Springs, I was reminded by street sign bearing your name, that I was overdue on thanks to you.

The occasion was a bombing raid briefing at 2300 hrs. on December 31, 1965, at Andersen AFB, Guam. My crew and I were going to miss your New Year's Day show because of this raid, and this was especially aggravating because we would also miss a chance to greet you when you landed late that night. Only after the briefing room was called to attention, and when Gen. Crumm and the senior officers came down the briefing room aisle, did we hear a familiar voice, "At ease men, smoke 'em if you got 'em!" You and Jerry Colona and Carol Baker put on a special show for the B-52 crews flying that night and into the next day. You appeared very tired and beat from the weeks of shows in the Far East, and squeezing in one more performance into the end of the USO Christmas Tour was above the call, but much appreciated.

Of the times of the Viet Nam war I remember this the most. God bless you Bob Hope and . . . . "Thanks for the memory."

(Then Captain) David C. Wellman

Mercer Island, Washington 98040"

A couple of months later I received the signed picture of Bob Hope inscribed "To David C. Wellman, Thanks for the Memory, My best, Bob Hope."

Tet 1966

It had to happen. After nearly a year of Arc Light deployment nobody had to pull a SIOP alert stint on Guam. During times of reduced bombing raids we pondered our assumed immunity. Were we exempt from such deleterious duty because there weren't enough nukes to make up a set of sorties? Didn't the poor bastards from Biggs AFB pretty much have the Far East covered with their reflex alert operation at Andersen with their 95th Bomb Wing B-52Bs? In their ivory tower planning had SAC forgotten about the 30-plus B-52Fs on Guam? Well, none of these turned out to be true.

But somebody back in SAC HQ in the crew harassment division was contemplating more compelling reasons. It took a Vietnamese New Year standdown to trigger an unbelievable switch to Cold War operations! Lt Col Jim Kleine, our popular 441st Bomb Squadron Operations Officer, delivered the summons to the crews that we would spend "Tet" on nuclear alert instead of enjoying a much needed few days off. Jim fit the bill for contacting crews hanging out at 9 p.m. in the O'Club swilling down reduced price drinks. Jim was a Dr. Pepper drinking Mormon tee-totaler! He herded the selected six or so crews into a separate briefing room where the Wing staff started the specialized crew briefings for the sorties. Those crews had drawn the short swizzle sticks. We were to understand that the threat of increased involvement of Red China in the Vietnam War was a reality and therefore we had the need for nuclear retaliation readiness.

The crews complained. (Don't they always?) We thought at the time that the mission was contrived. It may have been but, real or not, we didn't look forward to the requisite confinement to quarters. The weapon load was a single Mk-53, a relatively new 10 megaton monster that filled at least half of the B-52 bomb bay. It was a real Shanghai Buster! None of the crews had any familiarity with the bomb, having been associated with Mk-28's for their B-47 and B-52 years. We were given a quick primer (no pun intended) on the weapon, signed all the possession transfer papers and proceeded to the loaded aircraft in the middle of another rainy night. The MER's remained loaded with the M-117 750-pounders to ease the transition back to reality. Instead of jettisoning the 750s after takeoff in case of a launch, one of our squadron's brighter crewmembers suggested that they might be hauled all the way to the target and dropped to augment the nuke! After that weeklong duty, it was a welcome relief to get back to the war.

Most of the operational stuff that came down from HQ was usually well staffed and thought out. This didn't appear that way. "Crew harassment," however, continued to come in other innovative ways. For example: Many of the crewmembers took opportunities to place orders for Honda 90cc motorcycles through the always entrepreneurial KC-135 tanker boomer's stationed at Kadena. Those guys would find their way to Andersen after an Arc Light refueling to be strip alert tankers for returning B-52s. They took orders for the Hondas and other vehicles not available through the Andersen or Navy Base BX for delivery from the Kadena BX on some future tanker. Probably 75 or so were procured and found their way into the crew barracks area to be placed along the wall under the second story walkway. Life was good.

It had to be the same guy who thought up the "Tet" alert scheme that found his way to Andersen. Doing so gave him the opportunity to jump on an Arc Light mission to get combat pay and a month's tax exclusion. To justify his trip he did an inspection of the crew facilities. The unsightly Hondas parked along the barrack walls and out of the rain had to go – to a neat row along the driveway in the tropical climate of Guam - appropriately named "Hells Angels Parking! (Individual in picture is Capt Dick Hart, EW.)

Arc Light Hell's Angels Parking (Honda 90cc).

The Day the Spares Flew

In early 1966, we attended a briefing for a daytime mission to be flown by 27 aircraft (with two spares). The briefing was held the day before the actual flight. At the briefing was a representative of the Department of Agriculture who explained that the mission was a "trial" to test the concept of an approach to eliminating the jungle cover. TAC pilots would frequently complain of the difficulty they had getting visuals on the assigned target because of the canopy. The "target" was near the Cambodian border on a southwesterly heading. The target had apparently no military significance, but it was in an area west of Saigon where we had carpet bombed before. The Intelligence Officer briefed again the familiar warning to carefully avoid the Cambodian border, with a caveat that if we crossed it one of the half a dozen or MiG-15s in the Cambodian Air Force might take umbrage with our intrusion. We were also told that a couple "Dueces" from Bien Hoa would be along side.

The agriculture guy explained that this was to be part of an Operation "Ranch Hand." In the early morning hours in Vietnam, C-123's would fly over our "target" spraying the jungle canopy with a chemical substance that would dry up the trees. Several hours later, we would show up and bomb the sprayed area from 18-20,000 ft. using M-136 Incendiary bomblets loaded in an aerial mine shaped M-30 clamshell adapter. The "proof of concept" was to see if a fire could be started to clear the jungle canopy.

The 27 bombs were loaded internally. They weighed about 1,000 pounds each. The external pylons were downloaded. The two spare aircraft assumed their position on the east side of the runway ready to slip into the stream if there was an abort. If there were no aborts, the spares would follow the last airplane. The next day in mid-morning, the incendiary loaded B-52Fs took off and proceeded to the refueling area northwest of the Philippines. Each aircraft off loaded a token 20,000 pounds of fuel using three on one refueling tactics. (Shades of MCS!) The mission was proceeding to the usual ingress point of Cam Ranh Bay when a recall was directed and the airplanes returned to Andersen. Directions were given to land with, instead of dumping, the M-30's because of their incendiary nature and the restriction of weight on the lower station of the bomb bay racks for landing. It was done without incident. It turned out that a thunderstorm had swept through the "target" and washed off the defoliant. It was another incident of what turned out to be another "Download the bombs, upload the gas; pick your nose, scratch you ass" for the already overworked ground crews and bomb loaders.

Max Mihura – "Ranch Hand" Spare 2

Within the next couple of weeks the whole routine was executed again, except that time, my crew, E-26, which would usually be Red 1 when we flew, was positioned as Spare 1. Spare 2 was piloted by Max Mihura's crew from the Columbus, MS 454th Bomb Wing. On many of the missions, the Red 1 slot was either Mihura or Bradshaw - our 320th Bomb Wing. Although we were not an "S" Crew, we were almost exclusively picked to fly Red 1 with Col Wm. Cumisky, the 454th

Bomb Wing Commander, who was the senior ranking of the two Wing Commanders on Guam. That was because, giving credit where credit is due, of the recognized leadership and skills of our navigators, Frank Lenyo and Roland Larson. (Both Frank and Roland are now deceased)

There were no aborts, so both spares dutifully tagged along following the main strike force. Being the last to approach the target area, large cumulus buildups had already reached our bombing altitude aided by the heat caused by the fires. The defoliant had dried the canopy and the incendiary bomblets had done their thing.

It was unknown at the time of the mission, but the use of the 50-50 mixture of 2, 4, 5-T and 2, 4-D "Agent Orange" in Vietnam was going to become a source of much controversy.

This is the only B-52 mission I am aware of that was involved with the "Ranch Hand" defoliating effort.

John Bair in Southeast Asia.

Arc Light and MARS: Staying in Touch
John Bair

During the summer of 1971, I was a 1st Lt Electronic Warfare Officer in the Strategic Air Command. I had been assigned to the 449th Bomb Wing, 716th Bomb Squadron at Kincheloe AFB in the upper peninsula of Michigan (the "Deep North"). I was married while on my way to this assignment, and arrived there in early September, 1970. I was then sent Arc Light TDY to U-Tapao, Thailand, after being quickly checked out in my crew position on the B-52H. On the way over I was also checked out in the D-model at Castle AFB, California. From there we passed through Hawaii and on to Guam for some "in theater" training. This was air support for the ground war in Vietnam. I was married only eight months when I was sent overseas and consequently used every means available to stay in touch with my young wife, Helen. She had gone to stay with her parents back in Maryland while I was TDY.

To help pass the time while on Guam, my copilot Wayne Merritt and I decided to take a hike through the jungle to see if we could reach the beach. We walked across the golf course there on Andersen AFB and entered the under growth, down a rather steep bank and into the thick of it. After 20 minutes we could hear the ocean, but couldn't discern its direction due to the dense under growth. After some debate about getting lost, we decided it best to try to retrace our steps. Fortunately we came out of the jungle exactly where we had entered. I tried to convince my co it was due to my exceptional navigational skills! As an interesting side note, a Japanese soldier, Sergeant Shoichi Yokoi, holding out from WW II, was discovered near the Talofofo River on Jan 24, 1972. He had been living in a cave in a bamboo grove for the previous 28 years. This was just four months after we returned from Arc Light in September of 1971. Had we imagined there might be any surviving Japanese soldiers from WW II roaming the jungle we might have hesitated to venture on our little trek.

I used letters and cassette tapes to stay in touch with my wife and family, but the only "real time" communication was the MARS network as the phone system in Thailand was very unreliable. I remember unsuccessfully trying for over two hours to get an overseas operator on our first anniversary! MARS was the acronym for the Military Affiliate Radio Service. It consisted of a ham radio operator there on base, who would contact a ham radio operator back in the states, who would then use the phone land lines to connect to the desired party by collect call. Unfortunately since the radio portion of the transmission was "one way" at a time, you had to say "over" when you were finished speaking to let the ham operators know to switch from transmit to receive, and vice versa and to let the person you were speaking to know it was their turn to speak. The MARS network was the only reliable option. There were no internet or cell phones back then, and the clearest radio signal wasn't received until late at night there in Thailand (around 2 am.) This would be in the afternoon back on the East Coast, where my wife was staying.

We would always give an alternate phone number for the phone patch, in case the person we were trying to reach wasn't home. This was a "person to person" collect call from the West Coast. I always used my Mom and Dad's phone number as the alternate.

On one occasion I remember the radio signal was not very clear, and the West Coast ham operator informed me that they had a connection with the "alternate". So I began talking with someone who I thought was my mother - telling her of "practical" things, such as brassware and opal jewelry, which I was getting for her. But at the end of the conversation, after I said "over", back came this loud and clear "I love you, over!" Now my mother wasn't prone to verbal outbursts of affection, especially when the whole world could be listening in, as is the case with this type of transmission. But it wasn't until I was walking back to the crew trailer that I finally realized that Helen had gone to visit my Mom and Dad, and that I had just talked with her! In the pitch black dead of night I actually jumped up and kicked my heels together for joy.

To try to save on the phone bills, as a person to person collect call was rather expensive, I decided to have the west coast operator just call collect to "anyone". I wrote my wife and instructed her to inform the rest of her family where she was staying to not accept the call if she was not there. This backfired the first time I tried it. Her sister answered the phone. The operator asks: "You have a collect call for anyone from 1st Lt John Bair will you accept the charge?" Flustered she responded, "Uhh, she's not here." The operator not aware of our little scheme still queries, "Will you accept the call?" She finally gives in and says, "Uhh, OK; John she's not here!" I simply responded, "Then why did you accept the call?"

On another occasion, I remember sitting in the phone booth there at the MARS Radio Shack waiting for the radio reception to improve enough for a radio phone patch. To pass the time I thought I might as well try to get the overseas operator. For the next hour I sat in one of the booths and listened to "busy" signals on the phone, not really expecting to get anyone, but then to my surprise I raised the long distance operator in Bangkok. In a few minutes she had me connected to an operator on the West Coast. You can't imagine my elation as I relayed the phone number to her, nor my absolute horror as the phone line in the booth I was using was cut off by an airman working there in the facility just as the phone was ringing back on the East Coast. He said he didn't realize I was using the booth's land lines. I must say he was very apologetic and trembling quite a bit, realizing what he had just done to an officer.

Around half way through the tour my crew was scheduled for six days R&R. The aircraft commander, Capt Ruben Siebert and the radar navigator, Capt Al Schultz had decided to have their wives fly to Thailand and stay at a resort on the coast called Pattaya. That was a bit beyond my means so I coordinated with operations to see if I could hop a tanker returning CONUS to meet my wife in Hawaii. They agreed as long as I got back in time for my next scheduled sortie. It was about two months before our first anniversary so it was just like a second honeymoon. All went well until returning to Guam. The KC-135 had engine trouble that delayed my return to U-T by one day. I coordinated with the command post and was relieved to find that my crew was scheduled for DNIF cover, and there was no one sick! My biggest concern was that if there was a problem, it might jeopardize someone else trying to rendezvous in Hawaii.

I remember how truly separated and isolated from my wife and loved ones I felt, and how painfully slow time seemed to pass. The only letter I ever received from my father caught up with me when I first arrived at U-T. It was a simple father to son note of encouragement: "I know what you are going through." My dad was in the Navy stationed on Guam during WWII. He served in electronic surveillance in the Joint Communication Activity. He was drafted, leaving a young wife and two little girls behind. (I was born after the war was over.) His note really meant a lot to me, and it made me think of how truly isolated he must have felt. Mail was all he had to stay in touch and travel by ship was slow compared to jet.

As the tour slowly dragged on and the time remaining shortened, we EWs would leave notes on the main receiver's scope in grease pencil informing the next crew's EW how "short" we were - an artistic Kilroy barely able to reach the equipment saying "I'm sooo short I can't see the scope!"

Coming home was exhilarating. We were assigned to return a B-52D to McCoy AFB, Florida. We took off from Guam and flew non-stop, refueling in flight off Hawaii and then the West Coast. 26 hours later we landed. The hardest part of coming home was dealing with all the "booty" we brought back. I actually bought a Thai 28" bicycle and totally disassembled it to pack it in a box! What was I thinking?

It's truly incredible to me how long distance communication has changed since then. I routinely video chat with my sons and granddaughters half a continent away, to stay in touch and be a part of their life from a distance. But I thank God for the MARS network and all the people who cooperated together to help us keep in touch with our loved ones back then.

No Good Deed Goes Unpunished

Ken Brown

Hard work will get you to the top of the "heap," right? Yea! Maybe so in your world!

In my world I'm just a bit suspicious; let me explain.

I'm Ken Brown, Captain, USAF, and RN on crew E-04 out of Glasgow AFB on Arc Light duty at Andersen AFB from September of 1966 until April 1967. The deployment has been going well - so far. We have about 15 missions under our belts by Thanksgiving time carrying everything imaginable: CBUs, 500-pound streamline, 750-pound fat face on the externals (hell on fuel consumption), 1000-pound rusty traditional duds dug out of Dugway Proving Grounds. We drop them all, come home with clear bomb bays, and manage to keep our noses out of the limelight (excuse me: the trouble light).

Just prior to Thanksgiving the “wheels” dreamed up a plan whereby a number of '52s (nine) would be put on semi-alert at Andersen, loaded with 500s and ready to launch and be over the target (Nam) in under a day and a half. Actually the target would be determined and directed by a ground station. All we had to do was fly to a directed point, contact ground, and follow their directions. Guess who had the lead? Better yet, guess who always flew "tail-end-Charlie" on every prior mission? Yep! Us!

It's Thanksgiving Day and things are quiet at Andersen until just prior to 8am when word comes down to man the birds, start engines and launch. Somehow we receive instructions on refueling times, target times and location as well as IP to target info with ground contact info. Launch and refueling go off without a hitch and the bombing checklist gets underway. Our target area, by the way, is the mountain just north of Phan Rhang Air Base in Southern Vietnam, a favorite spot for the bad guys to toss rockets onto the base. At the pre-IP common point we contact ground and begin following their instructions. Halfway between the IP and target communication with ground gets broken up and the bomb run is aborted. "Proceed to your alternate target," were the last words from this ground station.

Suffice to say that I believe this whole action has been "rigged." In our strike folder is a flight plan and maps with lines from primary target to secondary. We relinquish lead claiming radar stabilization problems (back to tail-end-Charlie where we're comfortable). Anyway, north to Da Nang, west to the border, then southbound along the border where Groundspot directs us to our new targets somewhere west of Pleaku. Once clear of the target the remainder of the trip was uneventfully boring.

Back home the wheels got onto us at debriefing for not hitting the primary target but other than "duty crew," "taxi crew," "command post crew," and a host of other assignments we went back to being "tail-end-Charlie" until Christmas time rolled around with the usual bombing lull. Darned if we don't get selected along with a half-dozen other crews to fly an eight-hour "EWO training sortie" complete with a single bomb drop on the rock just south of Tinian Island and low level over-water and a pop-up on a volcanic island then a true-to-life celestial nav leg of two hours. Great Christmas present!

But there's more to this saga

A month or so later we're at base ops for the usual ice cream cone prior to flying and there's an R&R aircraft full of army types headed for Hawaii and their R&R. An army captain strikes up a conversation with me saying: "Would you like to hear an interesting story?"

"Sure! Go!"

We're army rangers out of Pleiku and we have been working the Cambodian border trying to keep the NVA from entering South Vietnam in force. Back in November we encountered a large force of NVA along the border. All attempts to defeat this force were unsuccessful - howitzers, TAC aircraft, etc. had virtually no effect. On Thanksgiving Day we were about to enjoy our turkey dinner when word came to the front "back off now!" The ground rumbled and shook like an earthquake as the B-52s dropped their ordinance. The next day we walked through a moonscape-like area unopposed. You guys are really great! He shook my hand and headed for his R&R.

I guess that's why we did what we did!

Capt Dave Wellman in Guam.

1965 - Initial B-52F Deployment to Andersen AFB

Dave Wellman

By the fall of 1964, I was getting used to my new unit, the 441st B-52 Squadron, 320th Bomb Wing at Mather AFB, California. Having come from a B-47 outfit in Ohio, I was looking forward to some warm-climate duty. Mather suited me fine. I was assigned to replace a long time copilot, Wil Harkins, who had recently been moved to an IP crew so he could get the stick time needed to upgrade to AC. I was assigned to Crew E-26 headed by Lt Col Alan Weddle and the guys were well adapted to working together and were quick to compare me to their old CP as crew members are apt to do. We respectfully learned from each other in the next couple of months.

By late fall, our wing's KC-135 tanker crews had already started "Young Tiger" refueling support for TAC fighter deployments to SEA and F-105 operations from Kadena. We bomber crews were starting to get an inkling that SAC was finally starting to take an interest in using B-52s with conventional weapons in Vietnam. The fleet's B-52Fs were modified by programs South Bay (28) and Sun Bath (46) to carry 24- M-117 750-pound GP bombs on external racks. Called "MER" racks (for Multiple Ejector Racks), they were mounted to an I-Beam which had been attached to the familiar AGM-28 "Hound Dog" pylon. The addition of these racks upped the B-52's bomb carrying capacity to 51when used with the nine standard vertical racks in the bomb bay.

Training

It was obvious that SAC was itching to get the "show on the road", as several plans and ops types showed up from Omaha in December to have us do a "proof of concept" flight demonstration. Routine combat crew training missions were suspended and about six or so aircraft were loaded up with 750's to fly the practice mission. Apparently someone at SAC was also worried that flying for four to six hours at low altitude was something to be verified, fuel consumption wise. On an overcast morning, we took off from Mather at one-minute intervals and flew trail at 1000 feet across northern Oakland Bay, out to the Farallon Islands off San Francisco, and then down the coast of California to a point abeam Vandenberg. After a two hour racetrack, the drop area was cleared by radar (I only saw one lonely sailboat in the Pacific the whole flight) and the bombs were dropped armed to explode upon smacking into the ocean. Can you imagine getting away with this assault on the environment today? We returned and turned in our special fuel consumption numbers, never to hear a word about the results.

Mission Planning

The primary target was the Phuc Yen Airfield, the major North Vietnamese MiG base northwest of Hanoi. The mission was to stage from Andersen AFB, Guam, with air refueling over the South China Sea and dropping down to low level for a three hour nighttime run-in at low level to the target. At the target, the familiar "pop-up" maneuver to bombing altitude was made to avoid fragmentation from the 27 – M-59, 1000-pound bombs which were in the bomb bay. The plan called for 30 bombers to be involved, each also loaded with 24 external 750-pound bombs. SAC was going to go to war in a big traditional way, obliterating strategic targets at night at low level!

BLU-3/B Testing

Also during this time, our crew was singled out to go first to Biggs AFB, El Paso, TX and then to Eglin AFB, Florida, for test drops of BLU-3/B "Pineapple" cluster bomblets. We called them "Shuttlecocks" for having the similar shape of a badminton birdie. This munition was loaded in two "Hayes Dispensers" which filled the bomb bay. Our crew was selected because of the excellent bombing team we had in Capt Hank Nielsen and Lt Roland Larson (both now deceased). Larson would go on to become the Chief of Bomb Nav at 15th AF and a lifelong friend.

Several days were spent dropping the "bomblets" at Whites Sands Missile Range and one of the ranges at Eglin. At Eglin, we were in a race track pattern with a B-47 piloted by Maj Max Mihura, a noted SAC pilot and later a B-52H test pilot for SRAM A missile launches. Both aircraft made numerous drops at 500', and at the end of their last run at 425K, Max pulled up into a beautiful climbing barrel roll, something I had always wanted to do when I flew B-47's but was restricted by having external tanks. For our B-52 drops we did not have external wing tanks, but still only flew at 350K. At Eglin, our radar failed, but we were able to work out a timing system to drop visually, which was appreciated by the Systems Command staff running the tests.

When the Hayes Dispensers eventually showed up on Guam in early '66, our crew was part of a mission of 30 B-52Fs who participated in a raid on one of the "jungle" targets. Flying at night, 10 bombers dropped 750's from normal high altitude; a half an hour later, our crew, along with nine others, flying at 8,000', dropped the BLU-3's ; then in another half hour, another 10 bombers dropped full 750-pound loads. It was quite a spectacular night as the "bomblets" went off like thousands of firecrackers, many of them delayed because of being hung up in the jungle canopy. They were still visible as we flew out of sight leaving Cam Ranh Bay.

Do We Deploy or Not?

During January of 1965, a waiting game was played until deployment. There were several false alarm "Samsonite Loading Drills" and the anticipation grew. There was an increased interest in what was going on in the war and the rumors mounted. Then on Wednesday evening of 10 February, we pre-flighted our previously loaded B-52's in the pouring rain. In one of my most memorable moments, I was under the open bomb bay with Larson and our gunner,

TSgt Mel Harmor; we were checking over the loading configuration of the 18 - 1000 -pound bombs. 1000-pound bombs were being carried because there was not a full complement available at Andersen. None were loaded on the lower stations because of landing restrictions with more than 1000-pound ordnance on that rack position.

In the dimly lit bomb bay, a voice came from one of the two shadowy figures in the forward part, "How's it going in there, men?" Thinking it was a couple of our Wing staff officers, I responded using not too polite language. This was not taken well by the questioner who turned out to be Maj Gen Archie Olds, the 15th AF Command who was inspecting the operation. After a short night, we got the call the next morning to report to the Squadron Building.

Deployment

On 11 February 1965, 15 B-52Fs and 15 KC-135s took off at one minute intervals from the non-optimum runway which was a tricky operation, considering eight of the bombers were on the alert parking tree leading to the optimum runway. An identical complement of 7th Bomb Wing bombers and tankers deployed from Carswell AFB, Texas. The initial B-52s would be "cleared to destination", while the tankers were cleared to "the Kadena Air Base." About mid-stream, the tower started clearing B-52s to "the Andersen AFB"; we were surprised at that call since all of that was supposedly classified. The skies were clear as families gathered in the picnic area on the base housing side of the field to see their husbands and fathers depart. My wife would later tell me that there many tears among the families as the airplanes streamed out of sight to the West towards San Francisco

We passed over the Golden Gate Bridge at about 20,000', an event that was reported in the San Francisco newspapers. Still climbing, we started our fan out to get on our random track for refueling. After identifying and hooking up with "our" tanker, we started to top off for the long Pacific crossing. Since the tankers were staging through Hickam, they had excess fuel for that particular leg. Shortly after commencing air refueling, an in-the-clear UHF transmission relayed through ATC advised that the B-52 in-flight gross weight limit was increased by 10,000 lbs. There was no turbulence, and the smooth touch of Al Weddle meant we could take advantage of that and then some. Just as well, because the B-52 fuel reserves at Guam turned out to be between 8,000 and 12,000 pounds for many aircraft.

We flew nearly over Midway and 150 miles or so north of Wake Island. In keeping with the classification of the flight, position reports were being coded and sent out. That was an arduous undertaking involving excessive HF radio chatter. Fortunately our ECM Officer, Lt Tom Robinson was a whiz at all things KAC-wise. After about 8 hours the crews were starting to request higher altitudes (all in code) in order to get more fuel efficiency. The empty MER racks were having an impact on the fuel consumption. Finally a message came back (in code of course!) to cease all classified transmissions. Crews worked out their own altitude problem; we ended up at 48,000' – it's a strange feeling to watch the airspeed indicator floating around 100 KIAS!

After 16 plus hours of chasing the sun, we landed at Andersen as dusk settled. The group from Carswell followed shortly as well as the tankers that were transporting more air and ground crews. Within days a couple of Fs from Castle AFB were added as spare aircraft.

The "plan" called for an immediate turnaround in preparation for the strike mission. It never happened. In fact, we didn't do a lot of anything for several weeks. Eventually daily briefings and crew training missions were set up. Pilot proficiency missions including close formation flying were flown. Crew training missions involved flying up to Okinawa to make radar bomb runs on the NIKE site near Naha and then flying down the Marianas, descending at Iwo Jima at dawn and eventually dropping a small load of bombs on Nafton Rock on Rota. Visually the island of Iwo Jima was hard to reconcile because my mind had a picture that was in black & white from WWII films. Mt. Suribachi and surrounding vegetation were then green.

Captains Hank Nielsen, Joe Robertson (KIA), and Wil Harkins. Joe Robertson was an aircraft commander of one of the two B-52's involved in the mid-air on 18 June 1965.

By June, the use of B-52s was determined to be in a counterinsurgency role against Viet Cong logistics centers and other base areas. On 18 June 1965, 30 B-52Fs were launched against such a target. Unfortunately, two B-52's crewed by my squadron were involved in a mid-air collision over the South China Sea, killing eight crewmembers, including my friend, Capt Joe Robertson. I placed a copy of the picture of Joe shown in this vignette at the Vietnam Wall, where all eight of the crewmembers' names are grouped in one section. These were the first of many SAC crewmembers who gave their lives in the service of their country during the Vietnam War.

Lt Roland Larson, Capt Dave Roeder, and Hank Nielsen. Nielsen and Larson were the Bomb-Nav team for Mather E-26. Dave Roeder, a copilot in the 441st Sq., later became one of the Iran hostages while with the Air Attaché in Tehran.

Chapter Five– Southwest Asia

Southwest Asia - [south-west] [ey-zhuh] – *noun* - the part of Asia that includes the Middle East.

Left to right are: Major Linwood "Mace" Mason, Capt Steve Jackson, and Capt Allan "Al" Cobb.

Desert Storm Combat

Allan F. Cobb

From the age of six, the B-52 was a part of my life. My father was an enlisted controller at the Griffiss AFB command post in 1966 and 1967 having just returned from a year in Vietnam. I still remember as we approached the Griffiss main gate on a cloudy summer afternoon, I saw my first B-52, a G-model flying around the pattern. I was about to start second grade. Little did I know I would amass over 2,700 hours flying them from various locations around the world.

Nearly all of those hours were aircraft and mission training in nearly all aspects of its employment from 1984 to 1992, spent in weekly (or more) flights, training, upgrading, instructing, and evaluating and more instructing. After five years at Carswell AFB, Texas, in all those roles, including repeated weeks on alert, tanker refuelings, arduous low levels, countless trips around the traffic pattern, exercises, the occasional deployment, and dreaded Operations Readiness Inspections, simulator sessions; I thought I had seen it all.

Such were the days as a B-52 crewdog in the 80s ... training, alert, inspections, exercises, and more training. In January of 1990 I went to Castle AFB, California, as an instructor, converting fuzzy, wide-eyed Undergraduate Pilot Training graduates and enthusiastic pilot upgrades into steely SAC combat warriors. Once in a while at both Carswell and Castle, I would be part of an exercise that involved dropping real bombs that actually exploded instead of transmitting an electronic tone or dropping a dummy shape that kicked up dirt. After a little research, I found out I even flew a few of the very same airplanes that were based at Griffiss when I was there as a kid, having over the years (decades) been transferred to Castle.

I knew many fellow crewmembers that for an entire career never saw conflict except for those few early on that were Vietnam veterans, flying missions from Andersen AFB Guam, U-Tapao AB Thailand, or Kadena AB, Okinawa. I felt I was destined to be one of the non-combatants ... or so it seemed. So it was a strange feeling on the night of February 22, 1991, to climb into that so familiar cockpit, the venue of countless training flights, and take it fully armed into enemy territory to use it for its intended purpose.

As a 328th BS instructor at Castle, I had some time off since student training was suspended while H-model crews were being "difference trained" to the G-model. Desert Storm had been ongoing for several weeks and the mission pace was such that more crews were needed, to be supplied from H-model bases. I had been notified by my flight commander that I was to deploy to Guam to be part of the 1500th BW(Provisional) along with other assorted 328th crewmembers (pilots and radar navigators) to ferry aircraft between Andersen AFB and Diego Garcia, British Indian Ocean Territory, since heavy maintenance was done at Andersen only. Each ferry was a one for one swap. It was perfect! Just hang out in Guam for a few days, ferry a fully-loaded B-52 with just us two pilots and a nav, then sit on the beach at Diego for a few days waiting for the next swap.

For lack of anything better to do, my nav (my friend Steve Jackson, a 328th instructor radar navigator) and I would observe some pre-takeoff briefings from the back of the room before the crews went out to fly their combat missions. Most of them were carpet bombing missions just across the border to hit the Iraqi Republican Guard. It was now three weeks into the war. Early in the morning of February 22nd, as I was in the back of the briefing room relaxing and listening, Major Linwood "Mace" Mason, one of the 4300 BS flight commanders, wandered in and spotted Steve and me. Out in the hallway, he flatly informed us we were in crew rest immediately for a 1930 pre-takeoff brief to fly a mission that night taking off at 2330. The moment I was anticipating arrived but was met with an odd mix of elation and dread. One problem was I had just slept the entire night before; getting rested for an all-night mission seemed impossible! But based on what I had seen so far, I reassured myself that it would be fairly low threat, probably across the border to hit the Guard. All I was told was that an additional mission tasking had come down and the squadron was hastily assembling any crewmembers that were available to fly; as a mission-ready instructor who happened to be there, I was in the line of sight and available! I would be flying with the new squadron commander, Lt Col Logan Kelly (who I knew at Castle as the chief of Stan/Eval; he had given checkrides to some of my students). Steve Jackson would be one of the navs. We would be number 3 in a 3-ship cell. The entire crew was made up of Castle AFB instructors; Judson Walls was the radar navigator and Gary Freeman was the gunner. Major Franklin, from Loring AFB, ME (our sole non-Castle crewmember) was our Electronic Warfare Officer (EW).

After a day of attempted naps and reading novels in my tent, I arrived at the briefing room where our mission was revealed; this was no skip across the border and hurry back mission! Our target was deep in Iraqi territory, the Taji Scud missile and armor reconstitution depot, about 15 miles north of Baghdad. This facility would be very heavily defended and the slide depicting the mission track confirmed it; numerous SAM rings, AAA locations, you name it! The planned mission track avoided most, but not all the ground threats. Previous attempts had been made to destroy this target; its destruction was essential to start the ground phase of the war; all Iraqi armor assets would then be isolated in the field. A total of nine B-52s were tasked against it; three from Jeddah, then our three from Diego Garcia followed by three more from Moron, Spain. Each formation would be 15 minutes apart. Each bomber was loaded with 45 500-lb. Mk-82s or (as in our case) 45 750-lb. Mk-117s carried internally on multiple-ejector racks and externally on heavy-stores adapter beams. Great, I thought... my first mission was going to be a doozy, and who knew what the outcome would be. All right, well here I am and so be it, whatever happens, happens!

We went through the usual motions, except this time there was a prayer from the chaplain, then our cell leader (from a Loring AFB crew, whose names I can't remember) conducted the standardized, but informal formation briefing on the plan and procedures for the evening.

We made our usual stops through life support, picking up lunches and jugs. The big difference now was that we were prepared for anything and everything; we had our survival vests, flak jackets, escape and evasion charts, and for me, an issued 9mm Beretta in the leg pocket of my flight suit. Even though smoking was officially banned in SAC aircraft since 1987, those of us who smoked (or recently had quit) were well-stocked with smoking materials as needed… this was war, after all.

All this was happening, just like all those training missions, but in the back of my mind I had no idea what I was in for. And I wondered how I would last the night having hardly any sleep through the day. But everything on the ground went well, no maintenance delays... and we departed into the dark Indian Ocean night on time at 11:30 pm local time. The EW, gunner and the Radar were veterans, having flown many missions up to this point. This was Logan's second mission. Gary Freeman, the gunner, was well-prepared. He'd brought along his tape player and plugged it into the interphone system; as I flew the departure in the darkness, Cheap Trick was singing "I Want You to Want Me" in our headsets. It definitely lessened the tension, at least for me!

We had two refuelings on our long journey northwest bound. Our first was from KC-10s also from Diego Garcia. We refueled to our max gross weight of 488,000 lbs., having to have the power at near max thrust just to stay on connected the boom behind the tanker. A few hours later we were met with another formation of tankers, this time KC-135s from Jeddah. By now we were in Saudi Arabian airspace and the festivities would soon begin. We ran the checklists to prepare for our entry into Iraqi airspace. Exterior lights were off, masks on, night vision goggles on. We maintained our formation position using the forward-looking infrared feature of the EVS (electro-optical viewing system) and the NVGs; electronic emissions were kept to the absolute minimum. We were cruising at a block altitude of 37,000 to 38,000 feet.

All was quiet and routine as our formation crossed the Indian Ocean, the Arabian Sea, then the dark desert below. An hour or so later we were crossing the border into Iraq. Eventually we were about 100 miles south of Baghdad. Then the fun began. In front of us, the sky was suddenly littered with tracers from AAA, some of them getting a little close, some more looping above our altitude. I announced to the crew, "AAA ahead." Judd Walls, our radar nav responded in a calm, humorous tone, "Guess they're not real happy about us being here."

Our route took us west of Baghdad over Tharthar Lake, then eastward where we would finally turn southwest for our bomb run. It was challenging to maintain our track, as we were encountering 150 knot winds at out altitude. Over the lake, we saw numerous bright flashes on the ground to the east; the bomb explosions from the cell ahead of us were pummeling the Taji depot... in fifteen minutes they would get a second round from us then more later; it had to be hell down there! I also couldn't help but notice the plumes from SAMs off in the distance in front of us. (But more on that later.) We were setting up for the bomb run, our interphone busy with checklists, our radio busy with calls from fighter support and air defense suppression. As number three in the cell we were at the top of our altitude block, with one mile spacing. So far so good as we approached the target... then our airplane shuddered slightly as we let loose our bomb load, all of them gone in a few seconds; the target was in a tightly confined area. We were in a left turn as soon as the bombs were gone, reversing course to get out. Major Franklin, our EW, called out a hit from an SA-3 acquisition radar, but was quickly gone. Then, while banked in the turn, we saw out the left window a bright flash on the ground, becoming a bright plume of light seemingly getting closer. Twenty seconds later another appeared. Both were stationary in the window. Logan turned to me and asked, "Those look like SAMs to you?" Nothing from the EW since it was apparent the missiles were coming at us unguided. "I think so!" I answered, whereupon Logan disconnected the autopilot, raised the nose, rolled out and then resumed and increased our bank angle. A few tense seconds elapsed as the plumes disappeared beneath us, not knowing what was to happen next. I stared out the window on my side and there was nothing. A few seconds later we felt relieved as the missiles seemed to pass harmlessly below.

Before we could relax, the EW called, "SA-6, target tracking!" Just then we heard over the radio on the strike frequency, "Coors 61, Magnum." That meant an F-4G Wild Weasel had just launched a High-Speed Anti-Radiation Missile (a HARM) that homed in on the SA-6 target tracking radar. Almost immediately the EW called, "SA-6 is down." Another restrained relief as we continued our egress from the area.

The next hour, we exited the same way we came, the route planned to avoid known ground threats. There was more AAA as we proceeded, but at last it finally subsided as we neared the Saudi border.

The next few moments were very strange for me, as the adrenaline from the last hour or so subsided. I suddenly started the laugh; the laugh turned to an uncontrollable giggle! How odd, a rational part of myself observed, accompanied with the thought that it was a good thing my mask was on and that my interphone wasn't in the hot mike position. I would have looked like an absolute lunatic!

The drone home was long with one more rendezvous with tankers; we let our gunner come up and fly the airplane while Logan and I took turns napping for the long voyage back to Diego Garcia. By then the sun had long risen and it was a new day. Compared to the previous night, it seemed so anti-climatic 16 hours from when we departed, to fly the ILS approach to the runway that beautiful afternoon to such a remote tropical paradise.

We saw the bomb damage assessment photos two days later; nine B-52s dropped a total of 405 bombs on the Taji depot that night. There was nothing recognizable save a dark, smoking hole. Almost immediately after our mission, the ground war began. I was to remain in Diego Garcia with the 4300th and be assigned to a crew as an aircraft commander to fly more missions per the standard schedule. Then in four days, the cease fire ensued. It was to be my one and only B-52 combat mission. I happened to be in the right place at the right time to be involved in the conflict, though many highly qualified B-52 crewmembers were denied the opportunity to be in combat in the aircraft that defined their Air Force career over the span of many years. What did they miss out on? It's hard to say, since many of our predecessors in the B-52 lived (and died) through Hell flying combat missions in the Vietnam War; Operation Desert Storm was by some measures not as intense or lengthy. But, combat it was, with all the risks it entailed, and I know in my heart that all of us who flew the B-52 then and now would have enthusiastically taken our place in defense of our great Nation.

I continued to fly the B-52 as a member of the 328th BS conventional flight, then finishing up as the squadron assistant pilot branch chief including more student training through the end of 1992. In the meantime, I interviewed for and was eventually assigned to the U-2 Dragonlady which I flew to the end of my Air Force career in 2004. But that's another story.

Don't Lower the Gear!

Dan Staley

From copilot to command evaluator I stayed in the BUFF over the years qualified in the B-52, D, F, G, and H versions. It was a fun ride. During that time I saw a lot of interesting reaction to some bizarre and stressful events. Some were scary and some were funny. The one thing I learned was to keep a sense of humor, nothing is as bad (or as good) as it seems at the moment. Here are a few recollections. They are not unique, I am sure anyone who flew can tell stories like this. I hope they bring back a memory or two. Enjoy.

It was common procedure, to check the bomb bay prior to landing whenever you dropped any iron. This process typically started far away from the destination landing base but was considered part of the arrival procedures. The aircraft descended to 10,000 feet or below and the forward crew compartment was depressurized. A crew member, usually the nav then crawled out the small door at the back of the lower compartment onto the alternator deck, around the front gear, and onto a 20 inch wide "cat walk" that ran down the right side of the air craft. The destination of this trip was the bomb bay. The crew member, wearing his helmet, carrying a 20 pound "portable" oxygen bottle, and grasping his US issued olive green goose neck flashlight (with the red filter) would nimbly crawl back to the bomb bay and make sure there were no bombs laying on the doors. This usually took 10 to 15 minutes depending on how agile the crew member was and how motivated he was to get it over with.

On this mission we were breaking in a new EW. This was his first assignment after school and we were his first crew. He was clearly well trained and very knowledgeable with a great "can do" attitude and the personality to match. He fit right in with the rest of the crew. He was motivated by an eagerness to impress so he volunteered to check the bomb bay on our next mission. As the aircraft commander I was not real keen on the idea but as the radar nav pointed out "he's got to learn sometime" and "besides, he did volunteer." We went over the procedures and the route, we even took him to an airplane and let him crawl from the crew deck to the bomb bay. No problems, no sweat, he had it down and was ready to go. The next sortie came soon.

The flight was uneventful and marred only be some bad weather in the air refueling route on the outbound leg. As we headed home the EW was clearly anticipating this new experience. We were still hundreds of miles out over the ocean and it was still night when the time came to start our descent down for the bomb bay check. The recover field was not yet visible in the darkness but was right off the nose; a small island in a large ocean. The copilot had just reviewed the approach plate and confirmed the landing direction. So down we came from 39,000 feet heading to 10,000 feet. As the copilot and I went through the checklist and got ready to depressurize the EW un-strapped and headed down stairs. The nav team was willing to let the young EW take the trip and were trying to talk it up over the intercom as the EW headed down to them. Now there is one thing I should tell you about our EW, he was very slight of build. Measuring five feet three inches tall and weighed 'nothing', he was by far the smallest crewmember you have ever seen. Small feet, small hands, small head, the helmet wobbled back and forth on top of his small shoulders. Wearing the helmet, carrying the oxygen bottle, holding the flashlight, and crawling back to the bomb bay in the cold and dark with the aircraft moving was going to be a challenge. We got to 10,000 feet and managed to stay in formation order. After checking with the gunner (who was now fast asleep) we depressurized the forward compartment, cracked the door, and sent our young hero on his way. As was the crew custom the copilot noted the time and hacked a watch while the nav stayed by the door in the lower compartment and waited. The EW was excited! The nav team was happy. I still wasn't keen on the idea, but off he went. Like Alice though the looking glass the EW was gone, through the hatch, over the alternator deck, and onto the cat walk into the night. Gone. Right after the EW disappeared into the hatch we hit some light chop, nothing bad but it did move us around a little.

And so we waited. At the end of 10 minutes the nav team stopped telling "E'dub" jokes and commented about the length of time he'd been gone. At the end of 15 minutes the nav was shinning his flashlight through the lower compartment door openly questioning what was taking so long. I was adding to the intercom chatter now and was expressing my concern, the nav correctly interpreted my comments as "get ready because you are going to go look for the EW". Soon came the command, "Nav, go find him." That was followed by some choice comments about the EW's linage, his likeness to certain body parts, and his mental aptitude. As the nav stepped through the entry door the copilot reminded us that we were only about 40 minutes out of our initial approach point for landing – time was a'wasting.

The rest of the formation, having successfully completed their checks, pressed on home. We on the other hand held our heading and stayed at 10,000. Looking off the nose I was aware of two significant things. The EW had been gone for a long time and there was lightening in the weather out front. Neither fact gave me comfort. The copilot had again hacked a second watch when the nav left to remind us of the time he was gone. The radar nav and I then engaged in a long discussion as to where the EW might be. The best guess was, not knowing where he was in the dark he had kept going to the 47th section. Well, it was a guess. The radar nav thought perhaps he had stepped down on the bomb doors and because of his size couldn't get back on the catwalk – a prospect that I didn't even want to consider. The copilot chimed in with the most plausible reason for the disappearance, "he's been abducted by aliens!" It was about that time when the nav reappeared in the entry door - alone.

"I went all the way to the jack-screw and back," he announced. "No EW." The radar nav few into a rage at that point. "When I find him I'll kill 'em," he declared as he headed to the entry door. Before the nav could say anything the radar nav was over the alternator deck and gone, his helmet and flashlight were hanging above his seat. I'll never know for sure what really happened next. We were over the field and we had to turn to avoid the weather at 12 o'clock. I started a right turn with the autopilot on being as gentle as I could, and as I rolled out of the turn the radar nav was coming back through the lower entry hatch pulling the EW behind him. The nav announced "They're back!" Now our radar nav was a six foot something inches lean jock who prided himself on physical fitness and I had no doubt that if he wanted he could injure our young EW. But my concern was unfounded and as he plugged back into the intercom he couldn't stop laughing. The nav was giving us a blow by blow of what was happening on the lower deck.

"I think he's going to wet himself, look at those eyes, they are as big as plates...what is he saying?"

The radar nav came on next and said, "All he'll say is 'Don't Lower The Gear!'" Seems' like during that first "light chop" our young volunteer got cold feet and tried to turn around. He slipped off the cat walk and fell down on the right front trucks and then onto the gear door. Because of his diminutive size he fit between the gear and the door, the harder he fought the more he slid under the gear.

There he came to rest lying on his back pinned between the gear and the door. The only thing he could think of was the copilot going over the approach plate – hence "Don't lower the gear!" Poor guy had been there the entire time. His flashlight was lost someplace else in the wheel well so he couldn't even signal, only bang on the gear door and yell "Don't lower the gear!"

When the RN crawled by in the dark he saw the light of the missing flashlight reflecting off the other side of the wheel well. He had to hang off the cat walk and grab the EW's hand and pull him out. The look on the youngster's face made the irate RN burst into a smile that didn't leave him for a long time. Back in the lower compartment, safe and sound, he was shaken but otherwise alright; mostly he was just happy to be upright again.

We quickly locked up the forward compartment and started into the Decent and Landing checklist. As the EW strapped back into his seat the copilot asked him to help him with the checklist – the next step was "gear down."

The "Word"

Dan Staley

The first night I went to Baghdad during Desert Storm I had more than a little apprehension to say the least. We had all seen the CNN feeds showing the SAMs and AAA and we knew it was not "light." Baghdad was the one place where the Weasels and EF-111s were not going to help up – the bad guys didn't aim, they just shot and there were a lot of things still being shot. Undirected fire is still fire and FLAK is FLAK – aimed or not.

I was in theater on "loan" from Headquarters SAC to the host Bomb Wing. We had deployed with the wing leading the bombers in. Ostensibly I was in charge of one of the wing's planning cells but I had taken to flying DNIF cover as a way to validate that our mission planning made sense and to maintain what little sanity I had left. I was flying as DNIF cover for a pilot who had been taken out of the schedule so I had only met the crew the day before the flight. My initial impression was that they were young. They seemed to know their stuff but had been the victim of some unfortunate in-flight bad luck. Part of my job was to see what the problem was.

Introductions completed, we talked generally about the next day's fight. At that point I knew where we were going but the crew would not find out until the mission briefing. They had the cockiness of youth and exhibited the juvenile behavior of someone going in harm's way and unable to stop it. The nav seemed to be the ring leader and chief "crew clown." He was a tall kid and everyone called him "Lurch." I made a mental note to keep a special eye on him the next day. Still, as we talked I was increasingly convinced they knew their jobs and the discussion ended on a good note. I told them to get a good night's sleep and I would see them in the morning.

The next morning we met at the mission briefing, they were pumped up but uneasy. Everybody talked at the same time and Lurch had them whipped into a loud laughing mob. It made it hard to see if it was pre-takeoff nerves or real behavior. The confidence I had seen the night before was gone; this might be a harder flight then I had assumed. Anyway, we headed to the aircraft. I smiled as we got off the bus and walked to the airplane. It was already pre-flighted; all we had to do was count the bomb pins, load our gear, and climb in. It was at that point that one of the iconic rituals of the bombers crews of Desert Storm occurred. They donned their "personal survival gear." Some of them had Glock 17s, some had Colt 45s, and some had multiple hand guns. All had some version of a shoulder holster. A knife in the boot was common and a few had their own bullet proof vests. My surrogate crew was no exception; they looked like a group of Mexican Bandito's climbing into the BUFF.

The takeoff and refueling were uneventful. The crew was performing well and Lurch was keeping everyone loose. As we turned north toward Iraq the crew's mood changed noticeably - no idle chatter, no smart comments ... clearly they were getting ready for business. The intercom had been quite for some time when Lurch came on .

"Pilot, Nav."

"Yes" I replied.

"We usually say a word about this point in the mission, is it okay with you?"

Not knowing what the "word" was but not wishing to change their routine I said "Go ahead." I fully expected to hear an Abbot and Costello routine or Porky Pig singing "*Fire*" but what came next was a lot more memorable. As we headed north across the border into Iraq that cold night Lurch cleared his throat a said:

"Do you not know?
Have you not heard?
The LORD is the everlasting God,
the Creator of the ends of the earth.
He will not grow tired or weary,
and his understanding no one can fathom.
He gives strength to the weary
and increases the power of the weak.
Even youths grow tired and weary,
and young men stumble and fall;
but those who hope in the LORD
will renew their strength.
They will soar on wings like eagles;
they will run and not grow weary,
they will walk and not be faint."

Isaiah 40:24-31

Amen.

Baghdad by Moonlight

Dan Staley

During Desert Storm B-52s initially operated out of three foreword bases: Jeddah in Saudi Arabia, Diego Garcia in the Indian Ocean, and Moron, Spain. Later they also operated out of RAF Fairford in the UK. It was not uncommon to join several cells from different bases into a large strike force, typically targeting Baghdad or Al Basrah. Basrah was always exciting because of the post-target turn required to avoid our friends in neighboring Iran. On this special night however we were going back to Baghdad.

The flight was normal until we were over eastern Saudi Arabia when the weather changed for the worse. The low clouds we were flying over suddenly came up to our altitude. That caused me a little concern because tactics called for turning off all our external lights as we pressed north. Leaving the upper rotating beacon on only caused a large red reflection in the clouds. So we went dark and just maintained station keeping. If they were going to track us by our radar emissions well, that would be okay.

We didn't have an escort but each designated strike package (B-52, Tornados, etc.) had a support package that contained EC aircraft (EF-111 or EA-6B), a set of Weasels (F-18, F-4), and a Mig CAP that were usually F-15s. They never flew with us but they were always around to cover us to the target area. Plus, there was always an AWACS (E-3) and JSTARS (E-8) looking over the battle field directing traffic and providing Command and Control as we went in.

On that night we were meeting two other cells, one flight from Moron, one from Jeddah, and two from Diego Garcia, and joining up northwest of Baghdad to put a 12-ship package across the target. The crew openly discussed the rendezvous with the other cells. They had taken off from their bases at different times based upon their location. The guys form Diego were first then Moron and lastly the Jeddah aircraft. We found the flight from Jeddah west of the target and started to maneuver to join them in trail. Clearly they understood what was happening because they turned to help us join and then turned back on course. The entire maneuver was done without one radio call.

Where, however, was the other flight? It was about that time, as we were due west of Baghdad, that we broke out into a clear bright moonlit night and the visibility was unlimited in the desert air. Off to the east we could see the lights of Baghdad, so I guess by then they figured we knew where they were so what the heck. A quick look around and we could see our two cells but no other. As a matter of fact, for having so many aircraft in the air at Desert Storm it was rare to see anyone else.

We were the only attack on Baghdad that night and I was a little surprised to see the city's lights on. But, maybe that meant the plan was working. By flying past the city heading north they would think we were going someplace else! We were going to sneak up on them and fool them by attacking from the north hoping to catch the air defenses off guard. It was a brilliant plan brought to life by some faceless, wingless staff officer at Offutt. But I digress. We were almost to the turn point where we needed to start our slow turn to the right and into the IP. As we went past the city, which was still way to my right, I was suddenly aware of explosions at altitude. As I watched in horror the AAA and SAMs became more and more active - somebody was clearly attacking. Then I heard the post-strike call of the missing flight. We had missed joining up with them and they had pressed on to the target without us. And, sadly to say, they were well ahead of us. The next day, during a rather animated and explicit phone call to the wing planners at DG, we would hear that there was a mix up in the timing and TOT but for then all we had in store were the Baghdad defenses and they were awake, ready, and mad.

Our only option was to press on with the attack. This time however the AAA was up and ready for us well prior to us rolling out wings level at the IP. Everything was firing, 37mm and 57mm explosions were everywhere, but the thing that got my attention the most were the 100mm and 130mm systems tracking us at altitude. There were a few tracking calls from the EW but most of what we were looking at were being fired optically since they learned early in the war what happens if they radiated to long. The lights below were so bright it was hard to pick out the SAMs. Most of them were low altitude SA-7s or 9s but right in the middle of the action came an SA-2. At 100 TG the aircraft altitude changed so much that I briefly thought the RN had released the bombs. "AAA," the copilot said coolly.

At 60 TG I announced "FCI centered" and the RN responded with "Moving the cross hairs;" "30 TG, centered, wings level;" "20," "doors;" "bombs away."

"Two SAMs three o'clock...tracking."

"Detonation one; detention two – we're clear."

"Pilot, post-target turn heading two-zero-zero."

"Two SAMs 11 o'clock, no threat."

"SAMs moving to 12 o'clock, cleared turn."

There were too many bad things in the air to really avoid everything, so we were just trying to miss the big stuff. The SAM call had put me in a steep bank to the right; descending would have just put us into more AAA, so I had to stay at altitude and try to go as fast as our BUFF would go. Once out of the target area we rejoined the cell and broke off with the other flight and headed home.

As we egressed out of the target area we went back into the clouds. After about 10 minutes we broke back into the clear and were between a cloud deck below us and a thin deck above. We could see stars through the clouds above. We were then heading west to exit hostile territory on the extreme western part of the country which was thought to be a safe area. However, at that point the EW's started making calls to the cell. At first it was "searching;" then there was a "tracking" call followed immediately by "break lock." Then, nothing.

Our EW had said it "looked like a possible SA-6 signal" but "wasn't strong and was gone." Everybody relaxed and settled back in their seats getting ready for the long trip home and coming to grips with the emotions of the last hour. It was "Miller Time"… almost.

As I looked out to the left, the clouds below lit up with defused light. Then out of the clouds came two SA-2 missiles. I broke into the missiles which were so close that I could see them clearly as they passed over the airplane and detonate above us. The rest of the cell scattered at my maneuver. The EW said over and over that he had "no threats." As we turned to rejoin the cell a new threat was evident. Out in front and slightly to our left were more defused lights in the clouds.

We braced for whatever was coming and I asked the EW for "threats."

"Clear," was the response.

As the missiles broke through the clouds we saw what they were. Iraq's vengeance weapon, two SCUD missiles that had been fired from the western desert toward Israel. We marked the position of the sites on the map for our intel briefing, sent a SATCOM message, and headed home. Just another day at the office.

Bruno Melton, second from the left, on "Day One" of Operation Iraqi Freedom.

Operation Iraqi Freedom: Day One

Bruno Melton

Pre-Launch

It was three days before "Day One" and everybody was excited because they were finally going to be notified who was flying on Day One. My crew consisted of me, a pilot with 4,300 hours, my copilot; a reservist major with 3,500 hours; my radar navigator, a reservist major with 4,000 hours; a navigator captain in my squadron with 1,00 hours; and a reservist electronic warfare officer (EWO) major with 4,000 hours. We possessed lots of experience from prior conflicts. Lots of experience period.

Finally the names were called out; I was number eight and was scheduled to lead the third formation of the strike package. Wow. The Group Commander told us the crews were selected by the experience of the radars on the crews and also determined by who else made up the crews. It felt pretty good.

The crew briefings and formation briefings would all be done the day prior to launch to leave time to study the details before the flight. Everyone was ready to get this started. All we needed was the code word on the schedule telling us what day and time we were to launch. I must have checked the schedule five times every day in the upcoming days looking for the code word.

Then the moment finally came. The code word was there. Times were set and we were excited! It was time to get the thing started and kick some ass! However, it was with mixed feelings – we had been briefed there would be lots of surface-to-air missiles (SAMs) and anti-aircraft artillery (AAA) being shot at us while in country. We had the training, and finally it was time to put all our knowledge into use. It was hard to get to sleep.

Day One – Launch

As show time was drawing closer, we were all on an adrenaline rush. Nobody talked much about how they were feeling, but we all had a job to do and that was our first priority. We knew our aircraft and the weapons we were going to carry. Now we just needed to know our flight plan, were we would deploy the weapons, where the threats were, and who would be part of our package for threat suppression.

Our aircraft, along with most of the others, were carrying eight internal Conventional Air Launched Cruise Missiles (CALCMs) and 12 Joint Direct Attack Munitions (JDAMs). The plan was timed to hit all the points on the ground at the same time. The big briefing slide had all the routes on the screen with one of them going north towards Baghdad. I thought to myself, "I wonder who has that one?" and as I looked closer to our route of flight, I realized that WE were the crew going north. I looked around the room and noticed the younger guys seemed relieved it wasn't them.

"That's OK," I thought. "Just as we practiced. Nothing different." I had already flown with my crew into Afghanistan and knew the quality of people I was flying with. I couldn't have asked for a better crew.

We looked at each other and smiled. The radar navigator said, "Piece of cake," and it was time to get to business. We all studied and signed out our escape and evasion packets, our flying packets, and our 9mms.

The ramp was busy. There were 11 bombers and 11 tankers - all ready to do their job. We also noticed three B-2 bombers also prepping for flight. They would be going in after us. Guess we would knock on the door, HARD, then they would follow using their stealth. We knew who the workhorses were.

We climbed on board, checked our parachutes, checked our equipment, and started our eight engines. I looked at my watch and then at the taxi schedule. Right on time, the first formation of three aircraft taxied to the runway. Later the second formation taxied. Finally it was our turn. "Time to go guys," I said. There was lots of chatter on the common frequency due to aircraft flying through some weather on the departure.

Getting close to Kuwait we put on our survival vests and prepped our aircraft for enemy territory - turning off lights, configuring our seats so they didn't transmit a beacon if we ejected, and just mentally preparing ourselves for what we had to do.

We entered Iraq airspace. We listened to the garbled radios and headed for our orbit/launch zone. In order to help out with communications with my wingmen, we used the inter-plane call sign of "Spider." It was easy to pick out among the other call signs since it was "my" call sign.

We turned north in our orbit and launched our missiles - then we got the JDAMs ready. By that time we found out the army had begun their move to the north and all the targets south of a certain geographic point were "not to be struck." I asked the navigator where that was in relation to our targets. He replied that it wasn't a problem, so we started north and tried to contact the Combat Air Operations Center (CAOC) for final approval. We got it five minutes shy of weapons release. We were going to release on some palace and surrounding bunkers. As we started the push, I looked out my left window and somewhere around the eight o'clock position I noticed a large fire on the ground. I was not sure if it was a missile launch or a target being destroyed.

We pressed on. Even though I was looking outside, I also had to look inside to make sure we were on course for the bomb run. We dropped all of our weapons, turned south and flew as fast as we could. I never saw any missiles or AAA being shot in our direction. Our EW was picking up signals but nothing was being directly targeted towards us, as much as we could tell.

As it turned out, we were only one of two crews to drop all of our JDAMs. It was a good day; we dropped all of our weapons on bad guys and didn't get shot. On the way home, the adrenaline was wearing off somewhat but returned as we got nearer the base because we knew we would be talking with the other crews about the mission.

The next several missions also involved weapons releases, but none of them compared to the excitement and unknown we faced that first night.

Get Home-ites

Dan Staley

As anyone who has been away from home for an extended length of time can attest, the thing you want to do most is "go home." This phenomenon is referred to as "get home-ites." Wednesday, the 20th of February, 1991, was just such a go home-ites day. I was at Moron AFB, Spain, which contained the Depot Level Maintenance operation for B-52s in the Southwest Asia Theater. The base was located 35 miles south west of Sevilla in the Andalusia province of Spain on the southern part of the Iberian Peninsula. It was day 36 of the air operations of Desert Storm and I, like everyone else, was ready for a break.

The Wing Commander had cornered me earlier and asked me to come by his office after the early morning launch to "talk." As any crewdog knows, that was not a good sign. I racked my brain to try to figure out what indiscretion I had committed but nothing came to mind. That sent me into a near panic as I had no way of knowing what the topic might be; therefore, I had to be ready for anything! Actually it wasn't that bad, as the head of a Wing Planning Cell I had a lot of opportunities to talk to the commander; both on professional and personal bases. Still, I wasn't real sure what the topic was going to be.

As I entered his office he was reading the morning reports. Putting down the stack of papers he turned to look at me and said with a large grin, "How would you like to go home for the weekend?" "Home?" Visions of my wife flashed through my mind. "Sure," was my quick reply. "Who do I have to kill?"

"Nothing that simple," he replied. "We need to get the Can Bird out of here."

"How does that get me home?"

The commander went on to explain the rather complex international agreement that covered the Status of Forces covering B-52 in Southwest Asia. He explained the shell game we played at the bases in Saudi Arabia and Moron counting tails of aircraft on the ramp, of course that meant the aircraft airborne were uncounted so we actually had more aircraft in theater than reported. That took us back to the Can Bird - it couldn't fly; it just sat there leaking fuel and taking up space.

It had to go and make room for an airplane that was operational and air worthy.

"Will you take it home Friday?" the commander asked again.

Without hesitation or reservation I responded, "Yes!"

You don't get to be a Wing Commander and not have any skills at manipulation and negotiation. By talking about the need to "go home" he played to what he knew I wanted to do. He just had to offer me the means; he knew I would jump at the chance. He also knew that by saying it the way he did, he had me hooked before I realized the real key phrase in the entire exchange, "air worthy." More on that later.

"This must be kept secret for now; get your nav. I have the copilot; no one else can go." "If word gets out everyone on the base will want to go, and you'll have a real problem."

"Plan to leave as soon as you can on Friday. Take the old bird home and bring me a new one back on Monday."

"Take the weekend at home. Sweet!"

I left the office and headed off to tell my nav. He wouldn't be hard to find at that hour of the day. As I slipped into his tent I heard the slow regular breathing of a man deep in REM sleep. Naturally, it was only 0900, he would be in bed until after noon.

Navigator: (noun), (1) a person who navigates; esp., one skilled or employed in plotting the course for a ship or aircraft. (2) a person who can sleep anywhere at any time and for a prolonged duration without guilt or embarrassment; usually associated with the phrase, "We there yet?"

Yep, that's my nav. In fairness he had been up most of the night planning the day's missions; so I guessed a little rest was okay. Eating and sleeping in the same day is a sign of weakness, and because I knew he had eaten breakfast, I had no reservation about waking him up with the good news.

"Want to go home for a few days?" I asked.

"Unless we are leaving now, tell me later," he replied. He threw a cold cup of coffee at me, rolled over and went back to sleep. Deprived of someone to be excited with, I went back to work in the planning cell.

Later on that afternoon the DO and Maintenance Commander took me over to the Can Bird for a reality check. As all fliers know, the

Can Bird is called that because everything that "can" be taken off of it has been. The Cannibalization Bird is the official maintenance nomenclature, and it is an aircraft that is sacrificed for usable spare parts for other operational aircraft, usually because it is not worth fixing in the first place. As I looked at the aircraft, I noted it did have all eight engines - a plus. The wings looked intact; however, there were several grease sheets on the ground under the wings to catch the leaking fuel. The tail was not in good shape, it had absorbed one too many SAMs. Most of the tires were there, and as I got closer I saw that one of the engine nacelles was empty.

"We'll hang an engine and get a tire," Delta was saying. "I don't think we can do much about the leaks except leave those tanks empty, and the tail, well, it goes that way." As bad as the outside was, it was soon apparent that those were the good points.

We entered the airplane and went to the pilot's seat. Crossing the upper deck I noted that all the circuit breakers were pull on both panels – something I had never seen - ALL of them pulled. Looking back, the EW's position was totally empty, just a seat, everything else was gone. The gunner's scope was there, but there was nothing behind it. It was just the cover. Moving to the pilot's seat, I noticed the hydraulic panel was intact, but I had already been told some of the standby pumps were gone. I had an ADI, and airspeed indicator, VVI, Altimeter, and Standby ADI. Everything else was gone. Oil pressure gages were all there as were the EPR gauges, EGT, no Fuel Flow; everything else was gone. I tested the light in the gear handle, at least it worked.

The B-52 carries 48,030 gallons of JP4, about half of it in its wings for stability and structural integrity. On this bird we had only the mains and they leaked – "just a little." The rest of the fuel was in the fuselage, but there again, it leaked from holes. Challenge one: we could not get enough fuel on it to one hop home. The B-52 crew compartment is normally pressurized to 7.45 psi, giving a comfortable 8,000' cabin altitude at FL350. Challenge two: the crew compartment could not be pressurized, too many holes to maintain any kind of normal cabin pressure.

"So let me get this right," I said. "We can't put enough fuel on it to get home?"

"Correct?"

"And we can't go high enough to burn the fuel we can put on efficiently?"

"Correct?

"Oh, I thought you wanted to do something that was hard."

Still, a weekend at home, I thought. "Hmmm. Okay; seems like a good idea," I said. And so we started.

First, because the heading was as stable as the fuel system we had an extra long oxygen hose installed behind the EW's seat. That way the nav could shoot the sextant and go up and down the ladder to his seat without having to carry an oxygen bottle around. We were going to be on oxygen this flight. Next we made sure we had two good sextants because we were going to have to take a fix every 15 minutes or so going across the pond. Next we made sure we had good radios, of which there was one. We installed another one in the RN position – there was nothing there anyway: it had all been taken out for spare parts. We pressurized the fuel system to make sure there were no fuel leaks in the crew compartment. Satisfied with the rest of the systems, we agreed it was ready to fly. We decided not to stress anything else as it probably didn't have too much left, and we didn't want to waste functionality with tests.

Now for the nav and copilot. The co was a Lt Col from the DO's staff. I was really never sure why he was going home; something about his family was all I got. I decided to leave that alone, as it was none of my business anyway. He was qualified in the BUFF but not combat ready, and certainly, flying wasn't his primary job anymore. Still, as we talked about the flight, it was obvious he knew the systems and had a good understanding of what we were up against. The nav and I had flown together for some time, and I knew he was going to have a cow about this flight. It had nothing to do with the serviceability of the airframe or the need to shoot the sextant all night. It was the pressurization that was going to send him over the edge, having to be on oxygen. He smoked and was used to smoking during those long flights. He wasn't going to like the nicotine withdrawals.

That night I met with the folks in the tent next to us. They were a KC-135 crew from the California Guard. Their job was to fly round-robin, base to base, moving spare parts and people around, and going back to the states. As a rule they went to Barksdale AFB to RON. Barksdale was also my home base so they would carry things back and forth. Going home they took letters and packages; coming back they brought in letters, packages, and food. No MREs for us, we ate Domino's Pizza and cooked two-inch thick steaks on our hibachi. When they would arrive at Barksdale they would call my wife and deliver any package or letter, in turn she would have them over to the house for a home-cooked meal. She would give them whatever she had to send, and they would return the next day. It was better than FedEx.

Anyway, we talked about the up-coming flight on Friday; they were also heading back to the states that day. Rather than going into Barksdale; however, they were going to Langley AFB; bad luck for us. Still, they could solve the fuel problems and get us across the pond. The next day we worked on the flight plan, but there wasn't enough time to organize a formation flight according to ICAO rules. We were going to have to go solo. The 135 pilot and I worked on that problem and came up with an alternative solution.

It was during the mission planning exercise that word got out about our flight, and, sure enough, people came to see if we needed any more seats filled. It broke my heart to say no because I knew everyone felt like me and wanted to go see family, but I had to say "no." Then came one of our flight surgeons; he was a young captain and had gotten married three days before we left for Desert Storm. If he could just ride home he, "would really appreciate it," was all he said. I talked to the Wing Commander about it and said I didn't care; he could ride in the IP seat; it might be good for him to see what sucking oxygen for 10 hours felt like. The Doc was in. When I told him, you would have thought it was Christmas. Only recently assigned to Barksdale he had not been in a bomber before; he'd logged all his time in a tanker. This was going to be a "treat" – right.

Friday came and we showed up at the airplane. Ominously, the fire truck was also there. He positioned himself off the left wing for engine start and followed us all the way to the hold line. Not a good sign. Because of the morning strike packages, we were not going to take off until after noon. Also, we wanted to do most of the over water navigation (shooting the sextant) when it was dark. Our KC-135 friends had planned to take off 15 minutes in front of us.

Flight plans filed, preflight complete, we started engines and taxied to the end of the runway - first the 135 and then the BUFF. ATC would not give us a formation clearance or allow us to takeoff together, but we had devised a way around that. After he had received his clearance sitting at the end of the runway, the 135 declared VFR. Cleared VFR, he took off and pressed for home. I had to wait 15 minutes for wake turbulence (hadn't they ever heard of MITO) and then was cleared for takeoff. Immediately after takeoff I declared VFR, changed my squawk and pressed on. Heading due south 10 minutes after takeoff, the tanker started a slow 270 degree climbing turn to the left. As he passed back through 360 turning to 270, I was executing a climbing right turn several thousand feet under him heading to 270 degrees also. By the time we were wings level I was tucked in the observation position below the tanker. Turning my IFF to OFF ensured no formation flying that night. It was also about the time that the tanker reactivated his IFR flight plan, and off we headed for the US of A. We leveled off at FL 250, west bound and down.

The nav started taking his three-star fixes and checking the heading system. We quickly determined that the heading system was a lost cause as it was only correct for a few minutes before precession drove it way off. We slaved our heading to the tankers heading and the nav started Rhumb Line navigation. We also went in for our first refueling topping off the tanks we had used during takeoff. Here the co was right on top of things; although, I have to say the airplane felt strange with so little fuel in the wings. After this first AR we decided to put a little fuel out in the wings for the next AR and see what happened. To our surprise we did not blow up. It was at that point that the first generator fell off line. We ran it isolated. The Doc was a little concerned with all the talk about fuel. He was also starting to wonder where it was that you walked around like on the tanker. It was going to be a long flight.

Over the Azores we descended down below 12,000 feet and dropped our masks for a while. The tanker was above us and in radio contact. With the break over, we started back up to the tanker to get more gas. The night was clear, although it was cold. The aircraft heater didn't work well, and we were afraid to go to Emergency as the uncontrolled hot air might ruin our entire day. We put on jackets and gloves and pressed on. The tanker pilots called to tell us they were going to stretch their legs and walk to the back of their aircraft; the Doc was green with envy. I reminded the Doc that we were a "combat" aircraft, and the 135 was "support." It made him feel better for a while until his bladder started acting up, and he made his way to the urinal. "This IS combat," is all he would say.

As the flight wore on, the Doc became more and more uncomfortable and the nav became grumpier. The Doc was well over six feet tall, so shoe-horned into the jump seat for all that time on oxygen was taking a toll. He finally went and stood on the ladder to get fully vertical. This resulted in him getting in the nav's way as the nav tried to shoot the sextant and keep up some kind of heading. In the end, the nicotine withdrawals got the better of him, and he snapped at the

Doc. The entire exchange ended with the declaration, "If man was meant to fly he'd have green baggy skin with zippers all over it." I was never sure how that fit into the conversation, but it seemed to lighten the mood.

Fuel was always the Achilles Heel of the flight; we could never get enough to go on our way. The mains were as full as we dared fill them and fuel fed from the body tanks through the mains to the engines. Inflight refueling was a challenge because of our gross weight and onloading capacity. We couldn't take on a lot of fuel at one time, and at the light gross weights, the BUFF was a handful. The co and I shared the refueling task but his lack of recent proficiency shined through here, so I did the lion's share of AR. That was okay. In a strange way, I was enjoying myself. The auto pilot didn't work, so the entire time we had to hand-fly. Then, of course, there was the hot spot on my head from the darn helmet. This was flying, not systems operation, no computer management stuff here. This was real Dash-One knowledge hands on flying.

The generators came off line after the third refueling, and since they looked stable, we left them alone and ran them isolated. It wasn't long after that before we experienced our first hydraulic failure; it was really okay because, as it turned out, the hydraulics were fine. The engine on the other was done. We left it shut down and reminded the Doc that the B-52 had eight engines and we had only lost one. Nearing the coast things were not getting any better. The tanker had given us the final offload and had to turn toward Langley; we pressed on west for home.

Before the tanker turned toward land, he activated our flight plan with ATC. We were going VOR-direct because that was the only navigation instrument we had. As we checked in on our good radio (the spare installed the RN positions would only work on one frequency) we heard the comforting sound of, "Washington Center, radar contact." Our IFF did work; what a relief. They knew we were a B-52 (the tanker had made that real clear), and they gave us a "welcome home." Even though it was dark as we crossed over Virginia, we could almost see home. Sometime around that point, we started developing some significant problems. The fuel we thought we had was gone. The electrical panel, which had run so stable, started to fail, and we had to shut down a generator and spread the load across two.

Our staff copilot started looking at alternate landing fields because we were not going to be able to get to Barksdale in the shape we were in and we had no fuel. "We can make Cincinnati or turn back to Seymour Johnson," was his advice over the intercom.

"Not gonna happen," said the nav. "Our pilot is going hoooome."

That was the end of the defeatist talk; I was going home if I had to flap my arms and carry that crate. The Doc was openly questioning his military insurance and wondering if he had included his new bride on his policy. With the help of ATC I established a phone patch with our Command Post, and told them I wanted a tanker to drag us home. As fate would have it, they had a tanker near us heading home from a training sortie. They diverted him, and we met over eastern Tennessee where we immediately got on the boom. He dragged us home refueling the entire way. It was impossible for us to keep fuel on the poor old bird. We pumped it from the tanker through the bomber to the engines. Our in-flight weight was around the 225,000 pounds minimum AR weight for the entire time. But we were flying, the aircraft was controllable, and most of all, the distance between us and home was always decreasing!

As we neared Barksdale the tanker checked in with the Command Post and got us our approach clearances. We started a decent on the boom and dropped off around 10,000 feet. As we pulled the throttles to idle and started to lower the gear, we lost our second engine. At first I thought it was fuel starvation, but a quick check showed that the fuel quantity was good - the engine had just failed. We left it alone and configured the aircraft for landing. My next concern was getting the gear and flaps down. Slowing down at that weight was a chore, but with the gear down, the speed and descent rates were manageable, and the flaps came right down. All we had left to do was land and stop and at our weight stopping wasn't going to be an issue. All I needed was one of the two - either brakes or drag chute.

Over the overrun, we lost another generator which caused things to black out for a second, but it didn't matter. The aircraft touched down about 1,500 feet from the threshold with airbrakes set to six. I slightly tapped on the brakes, and the chute deployed. We were home.

We made the first turn off and stopped straight ahead. "Let's just shut it down and let them tow it."

The nav was way ahead of the game and had the 781 complete. "Sign this," he said as I finished the checklist.

The co was talking on the radio to the Command Post and the acting Wing Commander was headed out to meet us, as were a host of cars, bread trucks, and fire trucks. As I exited the aircraft, the reception committee had already formed. The nav gave the forms to the first maintenance supervisor he saw. The co shook the acting Wing Commander's hand and talked to the other wing staff. The commander turned to me and, after shaking my hand, told me to go into crew rest; the new airplane was ready to launch in 13 hours! The Doc exited the aircraft and broke the ice by falling on the ground and proclaiming his love for all things "firm."

I shook the commander's hand real firmly, smiled at him, and told him I would be back Monday at 0800, and to have the aircraft's preflight complete and ready to go. With that, the nav lit his cigarette, and he and I turned and went to the SOF vehicle for a ride to Base Ops where the only real commander I had was waiting. She smelled great and felt even better.

On Monday morning we showed up to an already pre-flighted aircraft and headed back east. The nav was there and so was the Doc, but we had a new crew.

I smiled the entire flight back to Saudi Arabia.

Chapter Six – Bar Stories

Bar stories [bahr] [stohr-ee] – *noun* - a narration of an incident or a series of events or an example of these that is or may be narrated, as an anecdote, joke, etc. told at a counter or place where beverages, esp. liquors, are served to customers.

The Kiwi Colonel

C.K. McCusker

One morning on Guam after a particularly long mission, my bombardier, Spike ordered a beer and a chili donut. Spike, aka "Bombs," (since he refused to be called an RN saying he wasn't a nurse) invented chili donuts one morning when Gilligan's Island, a roach coach set up in the parking lot of bomber operations at Andersen, had run out of hot dogs. They went over so well that they had subsequently put "chili donuts" on the menu. Other crewdogs actually ordered them at times. This accomplishment was a source of great pride for Spike.

Very soon after Spike got his chili donut, a Kiwi colonel came up to Gilligan's and ordered a cup of coffee. A Kiwi is a "bird" without wings and we used the term to refer to all non-rated officers that we didn't like. Obviously, he was what we also called a F**king New Guy (FNG.) Only a FNG would drink Gilligan's coffee. We knew he'd probably be dead by the afternoon if he drank it.

I noticed he was staring at me while he waited for his coffee, which took time, since we were sure it was being made from discarded grounds and iguana piss. Finally, the Kiwi couldn't help himself; he had been staring at my fake nametag. There were shops over there where you could get any kind of patch you could think up made. I wore fake nametags attached with Velcro and positioned under my regular nametag on both my flight jacket and flight suit. They were created in honor of our "hero." I would pull off the fake tag before taking off on a combat mission and put it back on after we landed.

"Captain Yossarian!" the colonel barked at me while still focused on the tag.

The minute Spike heard "Yossarian" he spit the beer he had just started to swallow at least five feet, and quickly covered his mouth to avoid hysteria. I saw Spike's reaction out of the corner of my eye and almost lost it myself.

Captain Yossarian was the primary protagonist in Joe Heller's brilliant war novel, *Catch-22*. The book was a crewdog's "second" Bible. Everybody in SAC had read it. Everyone except, apparently, this one colonel.

The colonel went on. "Your sideburns exceed the AFR limit Captain Yossarian! In addition, your hair is unkempt and you are uncovered (no hat) outside. It is important in a combat zone that military decorum be maintained. Very important! Consider yourself on report captain."

I responded, almost sarcastically, "Yes sir. I sure do agree with that. I'll get them trimmed right away. I have no idea what possessed me that I would allow them to get out of hand. My hat is in this bag sir."

I opened the bag I was carrying and pulled out my helmet with the oxygen mask dangling and put it on.

The colonel was not amused.

"That is inappropriate captain. You should have appropriate head covering and your hair should be neat when in uniform."

Spike still had his hand over his mouth, but he was audibly whimpering with laughter, and the FNG couldn't see him since Spike was behind him.

I was tired, pissed, and amused by then. "Sir my hair is in disarray because my head has been stuffed inside this helmet for the last 15 hours. Would you like me to write down my name so you get in right on the report?"

"That's unnecessary Captain Yossarian. I have it memorized. Now take off that helmet."

I complied with the order.

Spike was holding his crotch and bouncing from side to side as if he feared he was going to piss his flight suit. Finally, he ran, whimpering, toward the awaiting crew bus. I noticed him running, still holding his crotch, and I bit my lower lip to keep from laughing.

When I responded, I was smiling. "Well sir, since you know my name, I need to be going. I'm tired and I need to get some rest. You can report me to the general if you like. He knows me and said that one more 'bad hair' report and I would get a note of insubordination put on my permanent record. It's permanent for about 18 months until I get out."

The colonel was wound up by then. "Wipe that smile away captain. This is a serious matter. It's nothing to laugh about!"

I had had enough. I couldn't believe that the guy would bother a B-52 crewmember that had been flying all night. "I would disagree, sir. Sorry, I have to catch that bus."

I turned away and almost jogged to the crew bus while biting my lip to avoid a laughing fit. The Kiwi colonel yelled after me, "Captain Yossarian! I did not dismiss you! You are not dismissed! Come back here immediately! You did not salute either! I am putting you on report for gross disrespect to a superior officer! Yossarian! COME BACK HERE! YOU ARE INSUB…."

The colonel's tirade faded as I ran onto the bus and the bus driver, who I later learned had been set up by Spike, slammed the door and took off. There were six other crewdogs on the bus and Spike had briefed them on the gist of the conversation. They were all laughing almost hysterically.

Spike commented between guffaws, "Can you imagine what will happen to that dufus when he puts one of the most famous characters in the United States Air Force and modern literature on report! I would do anything to be there when the wing staff reads it!"

When I told them I recommended that he report me directly to the general, the whole bus went crazy all over again. This incident would become legend.

And it did!

For a long time, all that one of us had to yell at the club or on the bus, or wherever, was "Captain Yossarian your f**king hair is out of limits!"

And then everyone would jump in.

"Godammitt, Captain Yossarian! Salute me when I'm proving I'm a moron!!"

"Captain Yossarian can you HEAR me!? Probably not since your hair is over your ears!"

"I didn't dismiss you Captain Yossarian! Come back here! You can't dismiss yourself Yossarian!'

The drunker it got at the club, the more the famous encounter would evolve.

"Captain Yossarian! You are disgusting!! You smell like sh*t and you look like crap!! What do you do that gets you so disgusting Captain?"

"Dig holes!"

"Where are you digging holes captain! Are these holes authorized?"

"SEA!"

"SEA? Captain Yossarian do you think I'm a f**king idiot!! You can't dig holes in the ocean!"

"Yes."

"Yes what Yossarian? Yes, you admit you can't dig holes in the ocean or, yes you think I'm an idiot?"

"ANSWER ME YOSSARIAN!"

By the way, Spike ended up with more Arc Light missions in SEA than any B-52 Bombardier.

NOTE: The story above was altered a bit and used in the novel *Yankee Air Pirates,* by C.K. McCusker. This must read book was first published in late 2010 by Deeds Publishing and is available from Amazon.com or www.yankeeairpirates.com

Ice for Christmas

Walter Want

We were on alert and in previous years every effort had been made to allow alert crews to be home either the night before Christmas or on Christmas day. We had been hit with an ice and snow storm and things were quite bleak and morale was really low when it was announced that no crew changes were to occur.

We were iced down, the airplanes were literally frozen to the ground, and there was no way to pull the intake covers off the airplanes. Most likely the blades in the engines were also frozen and would not even turn until defrosted by huge heaters. Forget firing the canisters and getting them airborne. Our families were also trapped in their housing by the huge storm raging outside. Something had to be done.

In our squadron was a cartoonist, a talented young captain and a new AC. During the night a SAC emblem with the 319th Bomb Wing Insignia and the words of "PEACE IS OUR PROFESSION" appeared on the end wall of the primary entry tube. It was truly a thing of beauty and evoked pride in every SAC alert crewmember on duty.

The old man showed up about 7:30 AM to share our Christmas joy and saw his present on the wall. He was really impressed until he rounded the corner to the right and discovered the rest of the story.

In large letters the words "Every XXXXing day of the year." completed the artwork. The painting was slightly and quickly modified and crews changed out at 11 AM and a very Merry Christmas was had by all the professional SAC crews.

Merry Christmas to all and Good Night.

Columbus Crew E-39.

Out of Bomber Experiences

Ray Schalk

In preparing for a previous *We Were Crewdogs* piece, I was shuffling through old Arc Light material and pictures from that period and lots of memories came back to me. Some of them were not directly related to the missions or flying, in that all of our time wasn't spent on those things. Different crews and support groups engaged in different activities to occupy their time. Within that, different individuals did as well. So I thought I'd put down a few of my own adventures with a twofold intention: first, these snippets might conjure up some similar memories for others who were there and secondly, they might give other readers a peek into some non-typical activities that occupied some crewdogs' non-duty related time. I hope you enjoy the read....

Crew E-39 on their Honda 90ccs.

Honda 90s

My two Southeast Asia tours occurred early in the conflict – '66 and '67. In those times my best diversion was my little motorcycle. While it offered a distraction from the missions, it also opened the door for many adventures that created numerous memories. Judging from the number of the little bikes in our compound's parking area, I would say it was an outlet for many other crewdogs as well. All six of us on Turner Crew E-39 got one. Considering that the island is only 30 miles long and four to 12 miles wide, there wasn't a lot of long distance riding; but it was something to do.

At Andersen AFB, an old C-97 was one of several planes used to "hop" around from Guam to different destinations in the SE Asia area. The one that made the trip to Japan (The Yokota Flyer) also served as a source for the bikes, most of which were Honda 90's. The cost was 90 dollars, plus 10 dollars for the crew chief. When we were off duty, the majority of our time was spent using these bikes. We rode them on quick and frequent trips running errands or to our own Tarague Beach. All day outings exploring the island were enabled by our Hondas. Here are some fond recollections.

My friend Joe Harris was the nav on Tom Tobin's crew and a proclaimed "rock hound." He found out that the University of Chicago was excavating sites where the original native Chamorro people lived. We became interested in learning about these ancient people and their history and the history of Guam in general. We spent a lot of time working at the "digs" and talking to the archaeologists. After they returned to the U.S., we continued down their path visiting other sites as well, searching for Spanish footprints and remnants of WWII locations. Guam has quite a history and we found ourselves doing a lot of beach combing too, finding and collecting many interesting artifacts. We donated the majority of the things we found to the small museum in downtown Agana.

Just riding for fun exposed one to many unique experiences. Guam's national bird was a flightless Ko'Ko, resembling a kiwi of the shoe polish can fame. The little fellows always seemed to wait until the last minute before dashing across the road. They were FAST! The trick was to "get one" with your foot from the bike – a kind of polo, I guess. Or maybe it was some kind of "fighter pilot" urge? I can't remember any of us ever being successful.

Coconut Crab

Before dark, cocoanut crabs would show up on the road edge and on the beaches, too. They were unique land crabs, which lived mainly on cocoanuts and fruit. Usually they were small and about the size of one's hand. They came in beautiful red to purple to blue colors and were considered a food delicacy by the Guamanians. They had one large claw and one smaller one, which were equipped with sharp edges. The smaller "cutter" claw could cut through a shower clod like tin snips. They were a force with which to be reckoned.

Once I found a rather large crab. I wanted to have it mounted and learned of an older fellow who had a farm up by the radar tower on Mt. Santa Rosa. What an interesting friend he turned out to be. As a young man, he had been a deep water coral diver. He had also served as a scout for the U.S. for the retaking of Guam. He told stories – WOW! He related that Japanese stragglers were still around and things went missing pretty often. Yeah, right! Years later, I learned that one Japanese soldier, Shoichi Yokoi, came out of the jungle in 1972, well after we were there. Should we have been looking over our shoulder?

As a sidebar, one of my more unpopular moves was when I brought back a load of fresh, melt-in-your-mouth pineapples from his farm. I shared them with the other crews and everybody thought they were great – until the next day. Those who were flying and had over-indulged had a hard flight, and "honey-bucket" duties were in order upon their return.

Now back to the cocoanut crab (which I did finally get mounted). I didn't have a place to keep the live crab; so the crew agreed to keep it in the shower. Our EW, Cecil Thonan, was not an outdoors kind of guy. Some will remember that he was a connoisseur of Gregorian Church music. Our prank was to wait around until he hit the shower and watch the fireworks. All had a great laugh, including Cecil, after he regained his wits.

E-39 and Andersen Trailer.

While the Hondas were the key to such things as exploring the island far and away, they served us best as transport to beach-related activities. Of course, there was a regular bus to our own Taraque Beach; but beyond that, the bikes gave us access to an unlimited number of beaches and locations, some of which rarely saw humans.

Snorkeling in these pristine Pacific waters in and around the beautiful reefs was unsurpassed. Water there is shallow enough so that limited gear was required – fins, goggles and snorkel, and a spear gun – all of which packed easily on the rack of the Hondas. The variety of colorful little reef fish was endless. They came big and small, as individuals, and in schools of thousands. Most were harmless but a few were dangerous.

Ray Schalk and his blue Honda 90cc.

We used to go to the famous Pirates Cove Bar and Restaurant (original liquor license number one – issued by the Naval Governor of Guam in the 1940's) located in Ipan, Guam. We would keep their aquarium stocked in saltwater fish. Also, Turkey Fish (Lion Fish) were poisonous inhabitants in and about all of the reefs. One had to be careful. Once I had a "near-miss brush" with a huge shark.

The west side of Guam is relatively reef-free and has access to the deep ocean. I was riding my bike and saw about a six-pound Parrot Fish in the water below. I dismounted and entered the water. The fish eluded me, staying just out of range along the steep coral wall. I was forced to keep returning for more air. All at once the sky darkened and an enormous shark passed above me. Talk about scared! I ascended the sharp coral by hand, shredding my fingertips – blood everywhere. I turned to look into the blue, blue water. Nothing! Must not have been my day for that shark.

Speaking of Parrot Fish, we ate well. We always had fish and lobster in our freezer. We would go snorkeling at night with the Filipino lifeguards at Tarague Beach. They preferred fish over lobster and traded us for them. We didn't complain.

The Black Baron Patch

Shortly after my March '66 arrival, word was posted that volunteers were needed to name and then design a 4233rd Bomb Wing patch to represent (at that time) the Ellsworth and Turner groups. Those of us who worked on the project had fun and the planning meetings helped pass the time. A very talented airman from the 484th came up with the final escutcheon. His name was Tom Murray. In the style of the preceding Black Knights (7th BW) and the Iron Dukes (45th/320th BW's) the name Black Barons was adopted. It was derived from our black-bellied B-52s and their crews. It had the AF shield with two mailed fists symbolizing the combined force of the two wings with one crushing a "V" and the other a "C." The background is red and green with red representing the Communist North and the green for South Vietnam. It had a horizontal line (representing the 17th parallel) separating the two. It then depicted a black B-52 delivering a bomb strike. I still have my patch.

Ham Radio Operators

Another activity that was popular was contacting (or trying to contact) loved ones back home. Of course this was a "pre-cell-phone era," and as I recall, we were only allowed one free landline call per tour. To everyone's delight, a group (mostly, if not all, EWs) stepped up. They requisitioned an empty top floor room in our barracks and set up an amateur radio operation (HAM radio.) These volunteers manned it almost around the clock. Crewmembers would come in, sit around and talk while their turn came to be "patched" through to home. It was so interesting, too, just to listen to the chatter as the call was bounced around the world. Did I mention that beer was available? Our room was just down the walkway and the flow of traffic passed our doors. That traffic was a good indicator of the popularity of this activity. Thanks to all you guys for helping the rest of us.

The Stereo Cabinets

Some will remember the popularity of the beautiful stereo cabinets made in Taiwan. My RN, Ken Palmer, and I wanted one, but we had burned our R&R going to Hong Kong. Well, a three-day break came along, and we got the idea to take "hops," signing on as crew members, to get over and back with one of them. What an adventure! Follow this: Guam to Yokoto, back to Okinawa, Okinawa to Ching Cheng Kong, and a train to Taiwan. We had dinner at the Royal Palace and the next day we found that there was a six-month waiting list to get one. Fortunately, there was one which was unclaimed. I liked it and bought it for 90 dollars, U.S.

Then the fun began. Remember, we were "hopping rides" with this big three-piece stereo about 3' X 2' X 10'. Route back: Loaded truck to Ching Cheng Kong, reloaded C-130 to NAS Okinawa; reloaded on truck back to Kadena AFB. There we loaded it into the next tanker scheduled to go to Andersen for alert. And that was that. After several days of anxiously waiting, our tanker arrived with the cabinet. We took it back to our room to keep until it was ready to go back to Albany, strapped in the bomb bay. A side bar to this is that mishaps occurred with these cabinets. One night we were awakened by the loud sound of a rifle shot. It turned out to be the top of Pee Wee Lambert's cabinet which had reacted to the humidity and split. That trip was an experience but I still have that stereo piece in my home to this very day, along with its obsolete equipment.

Cambodia Trip

While at U-Tapao, we got the idea to see some of Cambodia and, in particular, the area where sapphires were mined. Most of the crew were up for it so we picked a free day and left early for Sattahip. We dressed in civvies, and carried our .38 calibers in our bags. We made a deal with a baht bus driver and off we went.

This was a little risky, of course. We had been briefed about the price on the head of an American officer. Oh well! We were young! We experienced the pristine villages and countryside of Cambodia. It was quite the break from dusty, dirty, smelly Sattahip. The people were wonderful, but little English was spoken. I can still remember the faces of the children when the teacher brought them out of their bungalow schoolhouse to see these tall Caucasians walking down their street.

The village was all about sapphires. Everyone seemed involved. Like the beautiful displays of fruit on the vender's carts, the stones were displayed, by color, on paper plates in the shop windows. Some gems were being polished with machines; some people were sitting on the street polishing them by hand. We did buy some stones including one uncut directly from the mine. After a long day, we returned with memories of rubber plantations, beautiful jungles – birds, monkeys and all – and the most warm and hospitable people you would ever want to meet.

I think that as we get older, our memories soften some of our harsher experiences in life, and we retain the better ones in sharper focus. My "out of bomber" ones are still pretty clear and fun to remember. I hope yours are too.

Turner AFB Reprise

Derek Detjen

For years, a thought lingered in my mind to one day revisit our favorite old SAC base in Albany, Georgia, the location of so many memorable happenings in our early crewdog years. My all time favorite Air Force movie was the original *Twelve O'clock High* film, where Maj Stovall decides to revisit his old fictitious town of Archbury, in the United Kingdom. A couple of years after the end of WW II, he travels back to Archbury, and espies an old pirate Toby mug in a downtown antique shop, the very mug that rested on the mantel in the Officer's club, turned to face the wall whenever a strike mission was scheduled for the following day.

Inspired, Harvey Stovall rents an English bike (naturally!) and pedals out of town to visit the old 8th AAF B-17 bomber base. He leans his bike on the fence surrounding the now empty farmland, marked only by the remains of the old runway and control tower. His eyes begin to mist over until suddenly, the sounds of propeller driven engines being started bends the grass over and the movie segues back to the beginning of his unit's wartime experiences. I wondered from time to time if I might feel a similar nostalgia upon visiting old Turner Field? I had to find out.

In early 1967, the B-52D 484th Bomb Wing began to dismantle, and by July 1st, the base was turned over to the U.S. Navy, partly due to the increasing threat of Russian nuclear subs and our diminishing reaction time to launch the alert force in the possible start of WW III. The Navy held the base until the mid-70s, when it became the property of the Marine Corps Supply Center before finally closing as a military facility. Since the main runway had all the latest technology in ILS, GCA, VASI lights and sequential strobe lighting, it was assumed by many of us that the City of Albany would use it as their local airport, closing the rather dilapidated airport south of town.

Unfortunately, that did not happen, and the adverse economic impact on East Albany was severe. There were rumors that some local politicos didn't was President Jimmy Carter to be able to land there in Air Force One, and the airdrome property was sold to the Miller Brewing Company. A large brewery now sits where the main runway was once located. Many local businesses that catered to the large military and civil servant populations soon closed, and the surrounding residential areas have gone downhill noticeably. The old field is surrounded by vine-covered cyclone fencing, but glimpses of the old steel maintenance hangars, almost all abandoned now, are disappointing.

The base links (now the Flint River Golf Course) have expanded to 18 holes with a lovely clubhouse worthy of Augusta. The old base gym is still used for practice by the Albany Tech College teams and other civilians. The main base is now the home of the Albany Job Corps Training Center and is restricted from visitation. A stark reminder of the terrible flood of 1994 is the totally destroyed Navy Hospital that sits along the golf course road. Most of the base housing units survive and have become low-income residences, managed by the government.

In closing, I had no Harvey Stovall moments, just an overwhelming sense of sadness, seeing the remains of what had been a shining part of our nuclear deterrent force during General Lemay's and General Power's years as CINCSAC. We did our jobs as dedicated professionals never dropping a nuclear weapon in anger. Rest in peace, Turner Field, you defended your country admirably. It was a time when being a patriot wasn't politically incorrect, and we were all proud of you.

Here are a few other stories I recall from my BUFF days.

The Inverted Eagle

Probably no single phenomenon better illustrated to me the "us versus them" mentality of the crew force verses the wing staff than the incident of the inverted eagle. It began simply enough with a large, early American eagle plaque affixed to the wall of the officers' lounge, over the door leading to the alert aircraft. Some enterprising crewdog managed to unscrew the plaque and remount it on the wall upside down. The plaque was immediately remounted by Maj Harry Rhea, an ex-KC-135 pilot and the alert facility chief at the time. Predictably, the following day found the plaque inverted again. Maj Rhea must have complained to our DCO, one Col Willie Sonntag, known for his short and somewhat violent temper, for at the following Wednesday weekly Commander's Call, Col Sonntag informed all of us that the plaque would stay in its correct position or "heads would roll!" This threat had no effect at all, and in fact became a rallying force behind the continuing friction between the wing staff and the alert crews. The battle was about to be waged in earnest!

Soon afterwards, a giant effort was underway, cutting out large silhouettes of cardboard, full-colonel's eagles, which appeared inside of all the illuminated EXIT signs over all of the doors in the alert facility. It goes without saying that all those shadows were also inverted. At one morning briefing, Maj Rhea gave an impassioned plea to the crews to "please stop the inverted eagle stuff." The following morning, when Maj Rhea entered the briefing room to give the standard daily briefings covering weather in the enroute and forward areas, maintenance status, etc., he was surprised to find the entire alert force facing the back of the room. Someone had unbolted all of the theatre-style seating rows and remounted them backwards! Outraged, our erstwhile alert facility chief slammed the door behind him and found a dead sparrow, spray painted gold, hanging from the back of the door - upside down, of course! At that point, some of the crewdogs feared that we might've given Maj Rhea a one-way ticket to Milledgeville and the state mental facility!

On another occasion, Maj Rhea arrived one morning to find every piece of furniture and all of the lamps upside down in the facility. Next came the appearance of many inverted eagles on the facing wall of each urinal in the men's rooms. At that point, I believe the staff and Maj Rhea admitted that they had lost the battle of the inverted eagle, and little more was said about the bird!

One final, amusing anecdote about the bird's effectiveness. A few months later, Col Sonntag was sitting in the officer's lounge, looked at the bird and remarked to Col Rhea, "You know, maybe you should take the bird down and then re-bolt it to the wall." Harry replied, "Yes sir, but if we did, some crewmember would probably find a way to remove the cement block it's mounted on and reset the entire concrete block upside down!"

Surprisingly, neither of the officers, thinking all was well, noted that the eagle was already upside down! Final score: Crewdogs 1; Staff 0!

Old Training Schedules

One of Col Willie Sonntag's favorite statements was "the training schedule is set in concrete," whenever any crewmember brought up the subject of possibly making changes in said schedule. The crewdogs used to joke that the colonel was referring to the previous month's schedule, rather than the current one. During that time period, a major construction effort to redo the concrete paving of the ramp area in front of base operations was in progress. As the jack hammers decimated the concrete area, dump trucks made endless trips around the perimeter road at the north end of the runway, neatly stacking the slabs in an orderly fashion, like giant dominoes.

Knowing that the staff always used that same perimeter road enroute to the alert facility our own Capt Walt Costello, Turner AFB's answer to Bob Hope, had a great and devious plan in mind. Late one Tuesday evening, prior to the expected arrival of Col Sonntag for the weekly Commander's Call, Capt Costello made a huge sign out of a bed sheet with large lettering that proclaimed "NO DUMPING, OLD TRAINING SCHEDULES ONLY!" in bright red letters and hung it from the front facing of the concrete slabs!

The plan worked to perfection, for Col Sonntag's face when he entered the facility the following morning was a bright crimson; he was almost apoplectic and unable to speak as he entered the briefing room, to the smiling faces of the crewmembers! It goes without saying that the warning sign disappeared within hours of that briefing! No further words were necessary as the score now read: Crewdogs 2; Staff 0!

Shuffleboard Wax

Next to Capt Costello, probably our most creative mind belonged to a young navigator named Don Rosella. Possessed of a keen, creative and somewhat sarcastic mind, Lt Rosella was a particular aficionado of shuffleboard and the long, thin, polished surface and metal discs that were used to play this highly competitive game. If the surface was too slow, it was usually the fault of not having enough shuffleboard wax on the table. Lt Rosella began to bug Maj Rhea, "Hey Harry, when are you going to get us a new can of shuffleboard wax?"

The request was acknowledged, but time and again, the new can of wax never appeared. Finally, a few weekly tours later, Maj Rhea eagerly awaited the appearance of Don Rosella and his crew in the briefing room that Thursday morning. Spying Rosella in the hallway, Maj Rhea said excitedly "Hey, Lt Rosella, I finally got you a new can of shuffleboard wax," to which the lieutenant nonchalantly replied "Shuffleboard? I've never played shuffleboard." Maj Rhea went down the hall mumbling incoherently! Score: Crewdogs 3; Staff 0!

Thunderbird Flyover

Our aircraft commander, Lt Col Theodore Heiland was a most gifted pilot, as mentioned in an earlier volume of *WWCD*. A lover of big band music, particularly that of Ted Heath, Glenn Miller, and Bennie Moten. Ted also had a remarkable ability to consume peanut butter in large quantities. It was a joke on our crew that the only reason Ted used bread on his sandwiches was to keep the peanut butter from sticking to his fingers! Also possessed of a keen sense of humor, our pilot's room was the corner room at the end of a long, downstairs crew hallway, conveniently located next to the large loudspeaker that was used to notify crewdogs of pertinent happenings. Ted had a great recording of the Thunderbirds performing a low level pass over an air show crowd. Early one morning, he tiptoed out into the hallway with a long extension cord with alligator clips on one end, clipped the wires into the facility's PA system, and hit the play button, engulfing the entire downstairs area with the deafening roar of the jet engines. The wires were instantly removed, and Ted was in bed with the lights out as the melee outside his room ensued by many recently awakened crewdogs fearing a nearby aircraft accident. No one ever found out who did it! On another occasion, Ted performed the same maneuver again, that time with the audio of a Bob Richard's Wheatie's commercial, with a bowling ball rumbling down the alley and striking the pins with a violent sounding impact. Again, Ted feigned sleep and his tape recorder escaped discovery!

A Disappearing Room

Our copilot, Lt Dale Kuhns, was a literal mountain of a young man, having been both the heavyweight wrestler for the U.S. Military Academy at West Point, New York, and also the middle line-backer on Army's football team. Unfortunately for Dale, his years at the Point coincided with those of Roger Staubach and Joe Bellino at the Naval Academy, resulting in four consecutive losses to the Midshipmen. Dale was a force to be reckoned with, from doing his pushups with either arm to other feats of strength. On one occasion, he wrestled Capt Tom Johnson, who played fullback for the Washington Huskies, throwing him around like a ragdoll! We should've known better than to have played a practical joke on Dale, but Capt Benoit (RN), Lt McComb (nav) and I decided to remove all of his room furniture one evening while Dale was upstairs watching the late movie.

Dale's bed, dresser, desk and other personal effects were disassembled quickly and hidden elsewhere in downstairs rooms of the crew facility. Fearing the worst, the three of us retired to our crew room, put two end tables on the back wall and two beds end-to-end in front of them, a snug fit against our hallway door. With lights out, we awaited the fireworks. Soon afterwards, we heard Dale's reaction to his empty room and then he was beating on our door. "Alright you guys, I know you did it and you're in there; let me in!"

Silence, and then we heard him pushing against the door and the bed barricade until his fingers were visible and with a giant forearm, the beds went up like a drawbridge, snapping off the headboard of one of them as our enraged copilot entered the room! Our loyal RN George, jumped up on his bed, pointed his finger at us and said "I had nothing to do with it, they did it!" Dave and I began to laugh, which probably saved us as we could see Dale's facial expression and color return to normal. He ended up laughing also, but I made a mental note to NEVER mess with a 265lb, 6ft 5in Academy football player again!

The Noon Klaxon Check

Every SAC base had a particular time to locally ensure the operability of their alert klaxon horns, strategically positioned around the base to notify the crews of an impending alert. At Turner AFB, the time for the check was exactly 1200 hrs. local. Coincidentally, 1200 hrs. usually found a large number of crewdogs and maintenance support personnel having lunch in the alert facility dining hall. In the middle of a quiet lunch, the conversation would suddenly be interrupted by the blaring of the klaxon, causing all present to scan the room furtively like a bunch of combat squirrels, before the sound subsided and a normal meal continued. We knew that sooner or later, "the colonels," either locally or at Seymour-Johnson AFB, North Carolina, which was our division HQ, would initiate a real alert just to gauge the crew's reaction time.

One day the klaxon went off, everyone looked around knowingly, but it kept blaring and in an instant, 50 to 60 alert troops rose as one mass and all tried to get through a single exit door at the same time! One would've thought the planners would have at least made a double-door entrance into the dining room, but the exit still went off pretty well. It should go without saying that everyone's lunches were stone cold after completing the alert response, topping off our fuel, etc.

On another memorable occasion, that noontime check caused a rather hilarious reaction to a successfully accomplished prank. There had been an almost all-night bridge game the previous evening, and late the next morning found Lt Ed Noyallis, the EW on another crew, in a deep sleep in his bunk downstairs. Ed was famous for being a heavy sleeper, and a quick decision was made to totally rotate all the furniture in his room 180 degrees, including the bunk he was sleeping in. Ed slept through it all, as several of us tiptoed out of his room and shut the door. Soon the noon klaxon horn went off, and someone banged loudly on his door, proclaiming "C'mon Ed, it's an alert, wake up!" The sounds inside the room were like the proverbial "bull in the China shop" as Ed rolled out of bed, ran into the back wall, hit some furniture, and uttered some choice words before his eyes finally espied the shaft of light coming from under the real door. He knew he'd been had as he opened the door to the raucous laughter and applause of a bunch of crewdogs who had just enjoyed another noon klaxon check in a most unusual way!

Fill Her Up, Bud?

In every crew force, there were usually several pilots who simply believed that they were a cut above the other officer crew members on their crew and others. In truth, most navigators and electronic warfare officers had completed about an extra year of crew training before being ready to assume alert duty, many with advanced college degrees, but the plain fact that "it's a pilot's Air Force" has been true since time immemorial, and with some good reasoning behind it.

From time to time there was an undercurrent of "EW jokes," or "nav jokes," poking fun at the supposed inferiority of those non-pilot types one pulled alert or flew with. This was usually taken with tongue in cheek, but a few of the more notable offenders were dealt with in various ways. At Turner AFB, Capt Tom Tobin was the primary author of an unending stream of stupid EW or nav jokes. During one alert tour, a number of nav-rated folks had been running model trains on a neat HO layout located in an unused downstairs bedroom. Capt Tobin stuck his head in the door, asked if he could run them, and after some basic instructions, was soon activating and deactivating power from sections of track as he navigated his consist around the layout. Finally, he arose and declared, "Yep, my kids would really like this," refusing to acknowledge that he actually enjoyed it himself.

Retribution was at hand, for when Capt Tobin arrived at his VW bus in the alert parking lot later that day, he noted that the cover over his gas cap was not seated properly, and opened it to find an empty sugar bag encircling the cap. He was never quite sure if someone had actually put sugar in his gas, but the seed had been sown! The empty sugar bags appeared from time to time. On a few other occasions, Tobin's visits to the downstairs shower facilities coincided with most of the downstairs commodes being flushed simultaneously, giving him an unexpected rush of completely hot water. The EW and nav jokes seemed to subside after some of those incidents. Score: Non-pilots: 1; Pilots: 0!

Epilogue

As previously stated, the almost college fraternity house ambiance of our alert facility seemed to make the endless days, weeks, and years of family separation pass more rapidly and contributed immeasurably to the livability of our situation. The constant "us vs. them" mentality actually was a positive motivating force behind our wing's performance during those years. We noted that after Col Sonntag left, his constant testing and harassment of the crew force had resulted in our top performance on every higher headquarter's evaluation during his tenure. I have no doubt that it was due in no small part to "Col Willie's" modus operandi also!

Tommy Towery with M-17 bomb at old Carswell AFB.

Return to Carswell

Tommy Towery

(Editor's Note: In the infant stages of this volume I started working on this story. At the same time I also informed the other prospective contributors that I needed some new stories for a new book. I can't say great minds think alike, because I would then have to prove that both I and Derek Detjen possessed great minds; however, he and I both came upon the idea of doing similar stories, based upon the same theme. Since they both relate to different bases, and are unique in content and presentation, I decided to include both stories for your enjoyment. You read his Turner AFB story earlier. This is mine.)

Regardless of how we entered the Air Force and obtained our commissions, be it via OTS, ROTC, or the Air Force Academy, most of us fliers had the opportunity sometime early in our careers to study the movie *12 O'clock High*. It was used by the military to teach the dynamics of leadership and the different leadership styles of two different commanders of the same unit. In the movie Colonel Keith Davenport (played by Gary Merrill) had become too close to his men and was troubled by the high losses of his unit and is deemed unfit to continue his role as commander. Gregory Peck, as Brigadier General Frank Savage, takes command of the unit with the orders to get it in shape and his hard nose tactics makes him a tyrant in the eyes of the bomber crews he leads. During my 20-year career, I served under commanders of both types, at one time or another, and especially remember how I hated working for the tyrant type, which I vividly remember from my B-52 days at Carswell.

But there is another story hidden in the film's script I failed to recognize during my younger years, and have only come to appreciate now that I have retired from the service. It is the storyline accompanying the character of Maj Harvey Stovall, played by Dean Jagger. Stovall is not a flyer but is the Group Adjunct and in the opening of the movie, we find him returning to England in 1949, four years after VE-Day. Looking back at the first time I saw the movie, I did not realize that it was set and filmed so soon after the war ended. That is especially a strange feeling now as I look back and realize it has been over 40 years since the Vietnam War, my war, ended.

I always liked the character of Maj Stovall. He was just a man doing his job. He was not a war hero. He was a ground pounder because he was too old to fly, but still he did what he was assigned and had to do to make the unit's mission successful. *12 O'clock High* starts off with Dean Jagger finding a Toby mug in an antique shop in London and taking it back to his old World War II base of Archbury. The mug was like one his old unit used to announce when a bombing mission was scheduled. There was no real base named Archbury, but for the movie it was his base, and he walked around looking at the old buildings and overgrown runway as his mind takes him back to his war years.

In similar fashion to Maj Stovall, I found myself returning to the former site of Carswell AFB, Texas, last year via a Space-Available hop on a Navy C-40 aircraft from California. Unlike the return to Archbury, I did not find rundown and abandoned buildings, but still many things had changed on the base since I left the area in 1978. The most dramatic change was that it was no longer an Air Force base and no longer called Carswell AFB. With the end of the Cold War, the 1991 Base Realignment and Closure (BRAC) Commission recommended that Carswell AFB be closed by 1994. Later it was decided most of the installation should be converted into a Joint Reserve Base, retaining the Air Force Reserve presence that had historically been located at Carswell and also relocating the Navy Reserve, Marine Corps Reserve and Air National Guard flying units from nearby Naval Air Station

Dallas, also identified for closure. Today, the base is known as Naval Air Station Joint Reserve Base Fort Worth/Carswell Field and is operated by the United States Navy.

My wife and I arrived there late in the evening and got a room at the Navy Lodge, built near Lake Worth in an area between the old Officer's Club and the Alert Facility. It is a beautiful lodge, and its presence makes it hard to remember how the area looked when I was stationed there. It is an area where I used to take my daughter to feed the ducks when I was confined to the base while pulling alert duty nearby. It is very close to the place where the "family visitation center" for the forces on alert was located, where they visited with their families (or girl friends) while they were on their 7-day alert cycle and could not leave the base. The other memory that immediately popped into my brain, when I saw the location, was that it was on the same road where we used to have to run our 1.5 miles aerobics test each year to prove our physical strength. Those were hard times for those of us who didn't stay in as good of shape as we were expected. For many of us, it was the only time of the year when we actually ran – or did any organized exercise for that matter. Unlike many services, the Air Force did not maintain a rigorous physical activity schedule. Once a year we had to weigh in to prove that we were still within our weight limits, and run 1.5 miles to show that we could still do it.

I could never reflect upon the differences or memories of every building on the base, but there are a few places which stand out and must be noted. I could not get access to the old Officer's Club, which I remember being located at a prime spot overlooking the lake. We did not eat there very often back when I was a crewdog, but we had some good squadron and wing parties there. It was also the site of the weekly bingo games in which my crowd participated during my crewdog days. I do not know if they ever had a Stag Bar, but when I arrived in 1972 there was not one. I remember that we were not allowed to enter the building on one New Year's Eve when I was on alert because alcohol was being served. It seemed odd that the Air Force trusted us with an aircraft loaded with nuclear bombs, but did not trust us to not drink when we were at a place where others indulged.

On the recent visit, my wife and I had dinner at the bowling alley snack bar. There were few patrons that night, and there were more eaters than bowlers. We did bowl a lot back in the Seventies, but the small base lanes lost out to a larger commercial establishment off base when I bowled in the officer's mixed league back then. The on-base lanes were usually frequented when our alert duties restricted us to the base, and many of the crewdogs with families would meet with them at the bowling alley for recreation at night. It was also a place to grab a quick bite on mission planning days.

Near the bowling alley is the base theater. I cannot tell you the number of times I sat in its darkness, eating popcorn and drinking Cokes with my then-wife. There was a special row of seats at the back that was always reserved for Alert Crews and a special place in the parking lot to park our alert six-packs – always backed into the parking space so that we could have easy and quick access should the horn go off during the movie. Back then it did not cost an arm and a leg to see the movie or to get treats when we did. Even though we had free movies in the alert facility, our families could not join us there, so we met at the theater. Besides, some of the movies shown by the gunners in the alert facility would not have been appreciated by our wives and would have been totally inappropriate for our young children. So we took advantage of having a nicer theater on base. This particular one was the theater where we took my daughter Tiffany to see her first movie, *Family Plot*, in 1977 when she was about six weeks old. It was often so crowded that we were turned away. Today, it is only open on Friday, Saturday, and Sunday and though the price is still reasonable, it will never live up to the memories of its heyday.

We didn't have time to visit the Base Exchange, but it was still located where it was in the Seventies. It was difficult not to recall the ever-present sign stating "Alert Crews may go to the head of the line." As with the theater, we parked facing out of the parking spaces, and I have vivid memories of dropping whatever I intended to buy and dashing through shoppers to jump into the six-pack when the Klaxon sounded. I was stationed at Carswell when "the new BX" was built. At that time in the military it was almost the kiss of death for a base to get a new BX because it seemed every time a base built a new exchange the base was closed down. It was the site of many trips for the alert crews and one vivid memory was how my radar navigator had a check bounce because they did not think his signature was genuine so they returned the check he had personally signed. Even though it was not his fault it was returned, he had to pay the returned check fee and lost his check cashing privileges for a month. That also brings to mind how different it was back then, when no check could be cashed at the BX unless it had a Social Security number printed on it along with the writer's driver's license number. Now that's a real thing of the past.

The flying squadrons were housed in an old (I mean old) hangar that had been converted to offices and a large, auditorium style briefing room. The hangar building still stands but I did not go inside – too many dark memories of early morning briefings and no-notice emergency procedure tests. I did look for and could not find one sign I remembered. On the street side of the building was a door we usually used when we went across the street to the T-4 and T-10 simulators or the altitude chamber, where we did our annual altitude hypoxia and rapid decompression training. After the bombing had stopped in Southeast Asia the base painted a sign above the door to honor one of my fellow EW's, Wayne Wallingford. On his political website it states, "As a young lieutenant, Wallingford flew over 300 combat missions over Vietnam. As an electronic warfare officer, he flew seven missions during the famous eleven-day bombings of North Vietnam. He was the only electronic warfare officer to be awarded the Silver Star." The sign is gone, but not the memories of Linebacker II and my respect for his bravery. Seven out of eleven days!

The headquarters' building of my days still stands. It never looked like it should have been a headquarters, and always looked more like it was built as a dormitory to me. I still get a chill run up my back thinking back to the activities that took place in that building when I was a young crewmember. Even though I worked there when I was the S-01 Stand/Eval EW, I have more bad memories of the building than good. This was where crewdogs went when they had thrown a bad bomb, or busted a checkride, or was called to the commander's officer for any reason. It was not usually a friendly visit. One commander was extremely protective of his grass, and anyone under his command caught walking across the grass instead of using a sidewalk would be summoned to his office and placed on grass cutting detail for punishment. I would have been just as happy to have found that building replaced with a parking lot.

There are lots of airplanes (gate guards as they are called in the United Kingdom) around the base. Sadly, there is no B-52, but there is a small area dedicated to the Strategic Air Command's prior presence which has a M-17 thermonuclear bomb on display. It is certainly larger than the ones that may or may not have been on the alert aircraft.

The alert facility building is still standing but has been retrofitted for new purposes – and I imagine the occupants are sometimes haunted by the ghosts of former crewdogs who took great delight in finding unique ways to deal with being locked up there for 120 days a year – one week on and two or three weeks off during the Cold War nuclear alert requirements. I would like to think that every so often the "ghostly" Klaxon still somehow goes off in rainy weather, giving the new occupants a jolt.

Smith & Wesson Combat Masterpiece

As we were loading our suitcases back into the car a familiar sound filled my ears. It was not the sound of J-57 jet engines, but the loud cracks of gunfire coming from the base firing range. My thoughts immediately returned to the days of our annual aircrew qualification requirements for the .38 caliber pistol carried by crewdogs of the time. The pistol of choice for the aircrew at the time was the Smith & Wesson Combat Masterpiece. We were never issued pistols to keep, but only checked them out when they were needed. Such times included combat missions in Southeast Asia and when transporting classified mission materials while on nuclear alert. Each year we were scheduled to re-qualify our shooting skills with 50 practice rounds and 50 scored rounds shot at a man-sized paper target downrange. I first earned my marksmanship ribbon while I was at ROTC summer camp and had fired expert every year since and looked forward to getting to shoot. Others had a little more problem with their qualification. Later in my career a cost-saving initiative reduced the training requirements for previously qualified expert marksman and I did not get to fire as often or as many rounds as I did early in my crewdog days.

Another memory also emerged. Some staff weenie came up with the thought that during a civil insurrection, the alert crews in the alert facility should be issued M-16s and defend the planes and the weapons (that I can neither confirm nor deny may have been loaded into the planes) from the potential overrun of the gates by insurgents. The planners went so far as to take some of us to the firing range and let us qualify with M-16s before someone remembered that the over-riding SAC plan was for the alert crews to launch the aircraft and get them out of harm's way during those types of uprisings. No one had thought how we could defend the facility and planes when neither we nor the planes would be there in such a case. I was lucky to be in the first group that got to qualify, because the future plans for training were cancelled – with no explanation. It was the only time I ever got to fire an M-16 in my Air Force career.

I save the biggest, and what I think funniest, change in the base for the last.

The most interesting change I found at Carswell, was done to the old base hospital. This is the building where my daughter was born, along with the children of many of my fellow crewdogs. It was located outside the base gates, and therefore alert crews could not visit it – even to witness the births of their children. If you were on alert when the baby came you were out of luck. During the base makeover, the hospital was repurposed – converting it from a hospital to a prison for women. I found it amusing when I saw what it had become and for some reason my ex-wife does not find the humor I find today when I tell people that my daughter was born in a woman's prison in Texas.

Some people just don't have a sense of humor.

What was once Carswell AFB's hospital is now a federal prison.

Trials and Tribulations

Rusty Keller

My story of how I made my way to becoming a B-52 Fire Control Operator (gunner) was written and published in *We Were Crewdogs - V*. I now offer you a collection of amusing or interesting stories that I either witnessed, or was made aware of, during my 13 years or so (nine years as a Gunner and four years elsewhere) in the B-52 community.

Low-Level Escort

In an earlier volume of this series, my former EW told his story about a particularly strange thing that happened to our crew. I mentioned it briefly in my story in *Volume V*, and thought I'd now give my complete version, referring to some notes I made at the time for reference.

On December 9, 1975, during a low level mission in the North Dakota/Wyoming area, we experienced some very strong turbulence and were being banged around pretty bad. I had my radar on and noticed a strong band of video noise across the scope – while set at minimum range. I'd seen that before during tanker over-runs when my radar was painting the tanker above and slightly behind my airplane. On that day, I just assumed I had low level ground returns showing on the base of the scope. Suddenly, we were hit with a violent blast of turbulence, and my walk-around oxygen bottle came out of its mounting bracket and flew across the airplane, almost hitting the EW. My pilot, Capt Garret Cormandy, applied power, established a climb, and told the copilot, Page Wagner, to call center and report we were aborting low level and returning to Fairchild due to extreme turbulence. We flew home safely, and I thought that was that.

The next day, I was off duty and at home when I got a call to report to the squadron in uniform - which I did. Upon arrival, the rest of my entire crew was already there and were talking to two majors (neither of which I had ever seen) in class-A uniforms. They introduced themselves and started asking me questions. They asked me if I had seen this, or seen that; if I had noticed this, or that; if anything unusual had happened that I knew of. I said, "No," to all the questions, not bothering to mention the ground returns on the radar. My crew stood there grinning at me like they knew a big secret.

Finally, I'd had enough of the third degree, and asked, "What's this all about anyway." One of the majors said, "You were followed through low level by a UFO." I looked at him and replied, "Yeah, right!" He said, "No, it's true. It was tracked by the RBS radar, and several witnesses on the ground saw it. We took them as reliable witnesses: a forest ranger, and a county sheriff."

I asked him, "How far behind us was this thing supposed to be." He replied, "About 100 feet." He added, "When you guys aborted low level, and started a climb, whatever it was, went vertical and disappeared upward."

So, that's my UFO story. I don't know what it was, but a few years later, I considered it could have been an early test version of the F-117, or some other gee-whiz aircraft our government was experimenting with. I think the majors (knowing what it was all the time), might have simply been gathering facts on our ability to pick up this so called UFO on our radar and threat warning systems.

Alert Shower

While at Fairchild AFB, we stood nuclear alert at two places: Fairchild and at the old Glasgow AFB in Montana. I was a new gunner, on satellite alert at Glasgow one morning and taking a shower when we had an exercise. For whatever reason, I didn't hear the horn (Klaxton) blaring. I was just happily soaping up and showering when our CQ clerk opened the shower curtain, and told me what I was missing.

I rinsed the soap off, but didn't take the time to towel off. I jumped into my flightsuit, commando style, and ran to my airplane. Fortunately, we had a Christmas tree type alert facility, and our airplane was right outside. My pilot, Capt Garett Cormandy, knowing I was new, never said anything to me about it. He just smiled, and figured I understood I should make sure that it never happened again.

I learned my lesson - being wet, with nothing on but a flight suit and boots, is very cold during a Montana winter.

The Lunch Request

I went into the squadron building one day at Grand Forks after we'd been hit by an ORI and everyone was getting the pre-mission stuff out of the way. I was in the Wing Gunner's office, when he said to me, "Rusty, go find Major Carter and get his crew's lunch request." I was wondering why a major would have his crew's flight lunch request as that was normally the gunner's responsibility. I had never met, or heard of Major Carter at the time, and it took me quite a while to find him. The fact that he wasn't a major, but was actually a sergeant, with the first name of Major complicated my task considerably.

He was a young, black man, and a real nice guy. On April 11, 1983, he was on B-52G 58-0161 flying out of Robbins AFB, Georgia, when the crew reported variable weather conditions which included cloud obscured mountain ridges and peaks. While traveling about 320 knots, his aircraft impacted the south face of 7,050 foot Square Top Mountain, Utah, at approximately the 6,800 foot level, killing all seven aboard.

What Happens in Vegas....

We had a gunner at Grand Forks, who was ... should I say unique? I was an instructor gunner, and I hated getting him in our simulator. To do a complete equipment/modes check that would take any other gunner maybe 20 minutes, would take this guy the whole hour, and even then he wouldn't do it very well. Then I'd have to mark him unsat, and someone else would have to retrain him.

His pilot told me that during crew briefings, he could tell everyone else the bus time to meet at the squadron the next day, but then, he'd have to tell this gunner directly, almost as an order, to be there at such and such time or otherwise, the gunner wouldn't be there.

One time, the unit was expecting an ORI at any moment, and everyone had been briefed to stay in the local area, so they would be available. I stopped off at a bar in Grand Forks one evening, and ran into the poor pilot of that gunner. We visited for awhile, then he said to me, "Guess who I heard from today?"

I asked, "Who?"

He said, "Well, the phone rang, and I answered it. I heard, Sir, this is Sergeant XXXX, I thought I should call you, and let you know where I am."

"Okay," the pilot said. "Where are you?"

"I'm in Las Vegas," was the reply.

"I was pissed, and I told him to catch the first flight out of there and get his butt back here," the pilot told me.

They finally kicked him out of the gunnery field, and of all things sent him to the Military Police. Last I heard of him, he was driving around in his patrol car, caught himself speeding, and gave himself a ticket.

That guy was a trip.

My Mama Didn't Raise Me to Be a Gunner

I recall another gunner at Grand Forks who was a problem. This guy was new, right out of Castle, and decided he didn't want to be a gunner. He thought the only thing he could do to avoid it was to wash out of the field. He decided to do this by doing unqualified work during his initial training and evaluations in the B-52H. He actually had a good grasp of his duties, and responsibilities, but simply refused to perform them as was required of him.

Having figured out what he was up to, the Stan/Eval gunner, who would often fly with and qualify him, decided there wasn't anything this guy could do wrong enough on his final check-ride to get an unqualified rating. The first 30 minutes or so of the flight, the gunner was tasked to perform a Fire Control System (modes) check. He did his modes check, but did so in a very sub-standard manner. He normally would have never passed a check-ride with such a poor performance, but the evaluator intended to let him slide.

Immediately after completing his modes check however, he turned around to the evaluator, and asked on interphone with everyone on board hearing him, "Did you see how badly I screwed up that modes check?" The evaluator just stared back at him in consternation. The squadron commander was on the airplane that day, so the evaluator had no choice but to disqualify the gunner, thereby causing him to be removed from the gunnery field. I don't know where he went, but his days as a gunner were over.

High in the Tail

When I first went to B-52D models at Andersen AFB, Guam, I had some problems with motion sickness. There was a strong mixture of smells that would assault my nose during flight (oils, jet fuel, engine exhaust, and who knows what?) Added to that, was the sometimes violent motion I experienced in the tail. That combination caused me to use a puke bag far too often. I talked with the Flight Surgeon about it, and he prescribed some great drugs for me. He gave me uppers and downers, and told me to take one of each, at the same time, just prior to takeoff. The downers were supposed to relax me to the point where nothing would bother me. The uppers would bring me back up to a level at which I could fulfill my duties as a gunner, but not be bothered by motion sickness.

Let me tell you right now, they worked. I used them for about six flights, and the tail was comfortable, the colors in the sky were vivid and beautiful, and even my flight lunches tasted good! By the time I ran out of pills, I was used to the tail and was no longer bothered by what I experienced back there. Flying from that point on however was a bit more boring, and certainly not as pleasant as it was during my time of taking the little white pills.

The Gunner Astronaut

While flying as a gunner in the tail of B-52Ds out of Andersen AFB, Guam, I experienced negative Gs a couple of times. The first time, we were in pre-contact, moving to contact position for aerial refueling from a KC-135A tanker. Just before we got into contact, the tanker had an autopilot hard over. Both of my pilots (Maj Earl Kelly and 1Lt Dan Aronson) slammed the control wheels forward into the instrument panel and down went the nose. My inertial reel didn't lock on my parachute and it launched me upward into my canopy frame. The move didn't hurt me, but did crack my flight helmet. I smoked back then and my cigarettes, my lighter, gloves, and checklist were slowing revolving around in the zero-G gravity which I was experiencing. Eventually, we made a slow, controlled recovery from our dive. By then, I had recovered from my surprise, had figured out what happened, and was okay.

Major Kelly asked, "You okay back there Guns?" I replied, "I'm amused." He said, "You're what?" I repeated that I was amused. He then said, "Well, I'm not F$#%ing amused!" It was very funny to me. The very next flight, it happened again, only that time we suddenly pitched up into the tanker. I didn't hit the canopy then, but I still got to make like an astronaut for a few seconds.

They say the motion in the tail of the BUFF could get a little violent occasionally.

A Hole in his Scope

I don't recall the year this happened, but I do remember the gunner's name. Since I am not sure of the spelling, I'll pass on writing it here. I'll just call him by the nickname this incident earned him. He became known as, "One Shot."

He returned from a flight in a B-52D, and during the post-flight de-briefing with maintenance, the Fire Control debriefer asked him if there were any problems with the FCS system. He replied, 'The radar is shot."

The debriefer asked, "How so?"

One Shot said, "Shot. I accidently shot it with my .38."

He'd been sitting back there in the gunner's compartment playing with his .38 caliber revolver, and it went off accidently putting a bullet right through the middle of his radar scope. I don't how far the episode went afterwards, but I'm sure he probably had another debriefing with at least the Wing Gunner, if not the Wing Commander.

Jungle E&E

While at Andersen, my crew was tasked to go on an Escape and Evasion (E&E) exercise to the Philippines. We rode a tanker to Clark AFB, and were briefed on our mission. We would be playing the parts of shot-down B-52 crewmembers for the exercise. Maj Earl Kelly (P), and Capt Sid Black (RN), would go out into the ocean in life rafts. The rest of us; 1Lt Dan Aronson (CP), 1Lt Stan Albaugh (N), Capt Brad Moore (EW), and I, would be in the jungle trying to be found by the good guys, while trying to not be found by the bad guys.

Brad and I were together as one pair, and Stan and Dan were elsewhere as another. Brad and I lay there all day long (several hours) as motionless as possible in some tall grass. It was hot, and I don't know how many insect bites I had to ignore. Finally, darkness came, and we were able to get up and move, with our objective of reaching the top of a nearby mountain. It was about as dark as it could be, but we finally arrived at the top. Eventually, we established radio contact with some PJ's who were looking for us, and after a while, they joined up with us in the dark. After searching us thoroughly to make sure we were who they were looking for, they ignored us.

Sometime, later, Stan, and Dan showed up with the PJ's who had found them on a different mountain. Then, we all left our mountain top, headed for a pickup point somewhere in the distance. Those mountains had tiered rice paddies up and down all sides, so ... it's was walk across a paddy, jump off the edge, land on the one about three feet below, (usually falling down as it was too dark to see), then walk across and do it again, over and over. When we got to the bottom of the mountain, we proceeded up the next one, walking across the paddy, climbing up over the wall, and then walking across the next paddy, etc. That was repeated over, and over, all night. It was a long, long, tiring night.

My copilot was suffering from some sort of sickness - feeling bad, sweating, throwing up, and definitely shouldn't have been out there. He didn't complain though, and got through it all, but I bet he doesn't have any happy memories about it. Come to think about it, I don't either.

One time, we stopped at a small stream and filled our water bags with water. I dropped in some purification tablets, shook it up, and then drank some. It didn't taste very good, but seemed drinkable. I sipped on it occasionally for the rest of the night. Later, after the sun came up, I looked at the water in that bag, and almost gagged. It was brown, dirty looking, with all kinds of things floating in it. I hope to never drink anything that looks like that again.

After trekking all night, and into the next afternoon, an Air Force CH-53 helicopter finally came in and picked us up. We were taken back to Clark, debriefed, and released to quarters. A day or so later, we returned to Andersen.

I Fly F-15s

I retired from the USAF in 1986. In 1989 I went to work with the B-52G WST at Andersen AFB, Guam, as a console operator on the Pilot's, and the Defensive Team's Simulators. After a few months, the Bomb Wing was deactivated and all the B-52's left the island; however, we were still tasked to maintain the WST Simulator in working condition as per our government contract. The pilot's part of the WST was unclassified, so people started to come into the WST for sort of a show-and-tell.

I had a commercial pilot's license, and was the only employee on site who knew anything at all about flying airplanes. As a result, I ended up being the guy who would take most of the visitors on a B-52 flight. Normally, we'd just start out on a 10-mile final, and fly an ILS approach into Andersen AFB, and land. The simulator had full visuals, so we could see out the front windows, and it was very interesting. However, for me, it eventually got to feel sort of like I was running an arcade, and I got tired of doing all that flying for people who knew nothing about it.

One day I got an Australian guy in there with me. I asked him, "Do you fly?" He replied, "Yeah, I fly F-15's." I thought, "Great! A real pilot. We can have some fun for a change". I zapped us out 10 miles, took us off freeze, and let him fly. We were all over the place. That guy couldn't fly worth anything. We crashed, and then I put us back out and we tried it again. The second time, and the third time, I tried to assist him with some advice on what to watch for, and what to do. We crashed both times, so finally, I shut down the simulator, the pilot box lowered itself down off its hydraulics, and the exit ladder came down.

As we were exiting the simulator, I asked him, "You say you fly F-15's?

He said, "Yeah, F-15's, F-16's, F-4's. I have lots of airplanes on my computer."

Geeeeeeeezzzzz!!!

(Warmest regards to all B-52 Crewmembers and Maintenance Support Personnel past and present.)

Welcome to Roll Call!

Stephen "Mulligan" Henley

Disclaimer: The specifics of stories told at roll call are not normally told outside the squadron. I have explicit permission from each of the individuals cited to include their names and their "misdeeds."

I was asked to write a story for *We Were Crewdogs*, and currently missing from the volumes are stories of the daily life of crewdogs today. I think the best example of that would be to share some more recent traditions, such as "roll call."

Roll call is a weekly event (or as often as we can get together on a non-flying Friday) which is led by someone who is appointed as the "Mayor." It's not an official title of any kind. Roll call starts promptly at 1423 or whatever time aligns with the squadron's numerical designation. For example, the 69th Bomb Squadron starts at 1469 (aka 1509 - use your clock math to get this one or get the EW to pull out an E6B computer and do the math). The roll call begins with the Mayor announcing "Welcome to Roll Call!" immediately followed by a chorus of cheers.

The Mayor then takes attendance, a tradition evolved from World War II days when crews would return from a mission, and they would conduct a roll call in each of the barracks to see who made it back. Sometimes everyone was there; other times some guys were alone there. In any case, it starts with the reading of the call signs and a response from those in attendance. Sometimes guys will not respond with "Here," but with a humorous remark. If the individual called does not answer within about a half second, the Mayor will mark them down as "Shacked." Each "shack" is worth money to be paid to the bar (for the 23rd BS, its $2.30).

"Shredder?" "Here!"

"HEFOE?" "Here!"

"Bam-bam?" "Here!"

"Boom?" "Shugga, shugga, shugga!"

"Mulligan?" "Swing and a miss!"

"Rhino?"	"Here!"
"Sis-qo?"	"Banana Hammock!"
"Speed?"	"Here!"
"SWAT?"	"Anyone seen my pants?"
"Turret?"	"Here!"
"Warrant?"	"Drive like you stole it!"
"Veto?"	"Blame Canada!"

Upon completion of determining attendance of those with call signs, the Mayor will ask if there are any former Bomber Barons in the audience and will usually thank them for their attendance. After that, he will ask those without call signs to come up and perform a formation check-in. Each squadron calls their new guys different things. At the 23rd we call ours new guys FUNGUS's (F*** You, New Guy. You suck). Other squadrons call theirs FNGs (F***ing New Guys). The 69th BS calls theirs fledges.

So, they stand up in a line and start the check in just like they would in a radio check while in formation:

"Fungus check!" "2" "3" "4" "5" etc. Despite the fact that this is surprisingly simple, it is amazing how many times it gets screwed up…of course the heckling from the "named" aviators (those with a call sign) doesn't help. Sometimes a named person will jump in the middle of their "formation" and/or yell the wrong number. This is all just a little bit of low-grade hazing; not a big deal but a little bit of fun.

In order to make the gathering at least moderately educational, the new guys are asked to provide a useless fact and a bit of history. The history tidbit is usually related to the squadron or the B-52. My last roll call as a FUNGUS, I announced "Gene Roddenberry, the creator Star Trek, was a Bomber Baron." This is followed by a FUNGUS announcing a useless fact. Of course this fact can't be useful in any way. Someone once used the floor to announce "It is illegal in Bolivia to have sex with a woman if her mother is in the room." Immediately from the back of the room someone shouted "Useful! Try again!"

Next up: conformity check. Everyone is inspected to make sure they are wearing the proper Friday attire: T-shirt with a unit logo, nametag with their call sign on it, and squadron coin. Everyone is encouraged to "call out your neighbor!" Those missing one or more of those items are required to come up to the front, make a toast to the squadron, and drink a shot. These shots are usually chilled Jeremiah Weed, a thoroughly awful brand of whiskey which tastes like a combination of castor oil and turpentine; it burns all the way down and can be purchased at the Class Six for about a buck fifty. Though alcoholic shots are liberally encouraged, nonalcoholic shots are available. At no point did I ever witness anyone being forced to take an alcoholic shot.

There is a bit of "mob rule" involved and if anyone notices that a FUNGUS is going up to take a shot, they will announce, "When one FUNGUS drinks…" and the gathering will respond in unison "THEY ALL drink" and all FUNGUS's will take a shot with their brethren.

From that point in the festivities, the Mayor will go into to new business and old business: any announcements such as parties or friendly reminders. This is the last little bit of administrative business before the gathering starts to get a little more interesting.

First up is "Instant Justice and Laudable Acts" where you can call out anyone for anything. If called out on an Instant Justice, there is no chance for rebuttal, and you must take a shot. It should be duly noted that paybacks are hell. A squadron member will stand up, put his mug above his head and ask the Mayor for the floor:

"Mr. Mayor."

"Spaz, go ahead."

"I'd like to call out an IJ (instant justice) on Komrade. Our illustrious squadron physical fitness monitor failed to bring his elastic tape measure to perform waist measurements at the Physical Fitness Test."

Komrade went up to take a shot, toasted the squadron, and sat back down

"Mr. Mayor."

"Boom."

"Laudable Act for Rhino for being back to the pattern a full hour early on a long-duration mission."

<Polite applause.>

"Mr. Mayor."

"Boom again."

"Instant Justice for Rhino on the same sortie for being back to the pattern a full hour early on a long-duration mission and still managing to fly duration."

<Laughter from the masses.>

Rhino then went up to take a shot, toasted the squadron, and sat back down.

After that are the stories, where the bulk of time in Roll Call resides and where many call signs originate. Stories are required to contain a minimum of "10% truth". Please note that "He", "the", "it", "Minot", "B-52", "runway" etc. all count toward the 10% truth requirement…and we will usually round up to the nearest 10%. The stories traditionally start with a crewdog standing up in front of the squadron and starting with, "So there I was" and the crewdogs interrupt, in unison, "WHERE WERE YOU!?"

I would like to emphasize that ALL of these stories contain at least 10% truth…though none of them are 100% true.

"Mr. Mayor."

"FUNGUS Williams?" (Note: a new guy telling a story can be a gutsy move. For folks going into the B-52s, I highly recommend this tactic. It will command the respect of your peers.)

"So there I was."

"WHERE WERE YOU!?"

"I was over the Farallon Islands (a bombing range north of Guam) at 3,000 AGL. We just completed a dry pass and rolled in on our target preparing to drop a belly full of M117s on the range at low level. FCI was 0 and all systems were green. The Radar calls out '3 … 2 … 1 … weapons away…time of fall 13 seconds.'"

"At low level you can feel the sequential concussive blasts of the bombs as they go off. This day was no different and we felt the BUFF shudder. A few moments after shock arrival the aircraft commander called out, 'Crap! Roll wings level and climb!'"

"Immediately, we were all quite concerned. Personally, I was quite relieved to find that even approaching the age 30, all of my sweat glands seemed to be working extremely well. I placed my hands on the ejection handle and was ready to give the jet back to the taxpayers. We continued our climb until leveling off at 10,000 feet. The AC directed the copilot to double check the gauges and then asked the crew 'Did anyone else feel that?'"

"We asked 'What did it feel like?' (so to speak)."

"'It felt like the plane was shaking about 20 seconds after weapon release.'"

"'Sir, that was the bombs going off.'"

"'Oh.'"

"So," says FUNGUS Williams raising his mug in a toast "To Tarzan (our squadron commander!)" Tarzan then came up and took a shot.

"Mr Mayor!"

"Fridge?"

"So there I was."

"WHERE WERE YOU?!"

"I was in my office trying to knock out one of the 23 CBTs we have to do this week and I got an email from Ms Jones. For those who don't know, she's the lady in charge of our housekeeping staff here in Guam. I took the pleasure of printing this out so you all could hear what she had to say:"

"Dear Occupants of House 53, It was brought to my attention yesterday that housekeeping found one of the towels in the bathroom stained with an unknown substance. They contacted me and I came over to confirm the situation. Since we couldn't positively identify what the substance was, we contacted the Civil Engineering Squadron's environmental sciences unit to come over and check out the towel and the offending stain."

"They immediately quarantined the entire building and called in a biological HAZMAT team from Guam's rapid response unit and we also informed the Governor's staff. Three hours later, the team was released after the towel was sealed in a lead-lined, steel container."

"I don't know what you guys use these towels for, but please don't let this happen again. We are charging you $2.30 for towel replacement and we will consider filing criminal charges with Guam's finest should this situation re-occur. Hafa Adai!"

"So, to Mulligan for his inappropriate towel usage." And I took a shot, and it was the last time I used a towel from Guam to wipe off my checklist."

"Mr. Mayor."

"Dax?"

"So there I was."

"WHERE WERE YOU?!"

"I was going through the aircrew chemical warfare training class in the new classroom. We spent 2.3 hours meandering through a slow, painful, death-by-Powerpoint briefing. In this class they clearly describe how the decontamination process will run once you get out to the trainer. It's important that you actually go through the entire decontamination line both so that you know what you would have to do and so that those decontaminating you can practice it. In the decontamination process, they will strip you down just like they would in real life. That means your flightsuit is coming off. That's also why everyone wears shorts under their flightsuit on days when you're going to do this training. They also mention this quite clearly in the briefing and recommend that if you don't have a pair of shorts, at least get a pair of long underwear from the store room before you go."

"So, we get over to the trainer and I'm one of the first ones through the whole decontamination process. I put my flightsuit back on and then I hear one of the guys 'scrubbing' the aircrew say 'No way. You have got to be kidding.'"

"I look over just in time to see this poor Senior Airman stripping off the flightsuit of one of our fellow Bomber Barons. Underneath his flightsuit, he was wearing a T-Shirt and a...uh...<awkward pause>...a 'banana hammock.'"

<The crewdogs wince/groan and laugh.>

"Now being the good crewdog I am, I did what any self-respecting Bomber Baron would do: I whipped out my iPhone and took a few pictures."

At this point he turns on the TV behind him and goes through a series of slides showing the poor soul from various angles.

"So, to FUNGUS Collver-Horvath and his thong!"

<The crowd cheered.>

As FUNGUS Collver-Horvath walks up to take a shot, someone yells out, “When one FUNGUS drinks!”

“THEY ALL DRINK!” …and the rest of the FUNGUSs joined him for a shot.

As a side note, Collver-Horvath later obtained his call sign from this incident. Even though he wasn't wearing a thong as the storyteller claimed, the story stuck. A rapper called Sisqo made a song called "The Thong Song" and Collver-Horvath got the callsign Sis-qo.

"Any other stories?"

"Mr. Mayor."

"Tapper?"

Tapper walks up to the front of the room and starts "So I'm out on the flightline..."

Someone in the front interrupts with a muttered, "Where were you?"

"Oh, sorry. “So there I was.”

“WHERE WERE YOU?!”

"I was in hangar seven doing a weapons preflight with several noobs. First of all, even though it's indoors, it is balls cold. It's probably negative 25 and there's a slight breeze because they can't get the doors completely closed. So all of us are bundled up in about 23 layers of clothes. While we all look like lost Eskimos we're trying to get the preflight done as rapidly as possible. One individual is going through the weapons in the bomb bay and keeps removing her sleeve-length mittens and gloves to check every weapon's settings.”

“That's potentially kind of dangerous in this weather, so, I stopped her and asked why she kept doing that. Her response was, ‘I can't check the weapons with the mittens on’. Well yes, but why remove your gloves too?”

“‘It's because I have...these...man-hands.’”

“So, to FUNGUS Hester and her ‘man-hands.’”

The crowd laughs as FUNGUS Hester (who eventually got the callsign Maha) walks up to take a shot. Then someone shouts out “When one FUNGUS drinks!”

“THEY ALL DRINK!”

The mayor then pipes up "Are there any more stories? Going once, twice, stories are closed."

When all the stories are done, the Mayor calls up the person who had the best-voted story the week prior. The 'winner' each week is given the responsibility of holding onto the squadron mascot until the next roll call. Our mascot, a little stuffed donkey named Punch, has his own log book and has taken many flights as an unqualified crewmember. He even has some PIC time. The winner is expected to tell a little bit about what Punch did during the week under his supervision. This little burro's escapades are usually grossly exaggerated. During a week I was responsible for Punch, this jackass was arrested for drunk and disorderly conduct on an airliner, got lei'd, was thrown off a golf course for improperly using a ball washer, and joined the mile-high club (photographic evidence provided).

Punch the mascot.

With Punch's weekly escapes told, the previous winner returns Punch to the Mayor and the Mayor calls up everyone who had a story told about them. The masses then vote on what story was the best as the

Mayor holds Punch above their head like the Sword of Damocles and gives the story a humorous title.

“First, FUNGUS Hester with ‘I've got man hands and I know how to use them.’”

“Yeah!”

“Next is Mulligan with ‘Hey, sometimes a guy gets lonely.’”

<Crowd murmers.>

“Mulligan, go sit down.”

“FUNGUS Co-Ho with 'Swing low sweet chariot.’”

<Crowd mumbles.>

“All those voting for Tarzan and ‘What the hell was that!?’”

“YEAH!”

“Looks like Tarzan's the winner!”

The winner then takes Punch and is responsible for him the rest of the week.

Finally, if anyone is leaving the squadron, they will be called up to give the last toast of the evening with the squadron toast. If not, usually the winner of Punch will close us out:

“Fight’s on!” “Fight’s on!”

“Fight’s on!” “Fight’s on!”

“Fight’s on!” “Fight’s on!”

“Bomber Barons!” “Fly! Fight! Win! HOOAH!”

This depiction of roll call is centered on my experience in the 23rd BS; however every B-52 squadron has a similar format for roll call. This submission for *We Were Crewdogs* was meant to give the reader a bit of the flavor of the life of crewdogs today and to share the unique camaraderie that has been experienced by crewdogs throughout the many combat and peacetime years of the venerable B-52 BUFF.

The Stag Bar – An Exposé

Tommy Towery

There are some things in life that are probably best kept as secrets – things like a lodge's secret handshake or the thermonuclear war go-codes. The story of the stag bar is probably one of those things as well, but to a historian who is also a writer it is a story that cannot be kept hidden, even at the cost of fear of reprisals. It needs to be documented before all those who can remember it lose the ability to do so, either to the grasp of the Grim Reaper or to the usually progressive deterioration of intellectual functions such as memory.

Years ago, in September of 1991, the Navy got its profession reputation's nose bloodied when their behind the scenes activities at the 35th Annual Tailhook Association Symposium were made public. Although the Air Force skated on that investigation, they were not the clean cut church-goers that some wanted people to believe. But it was a different era and things that are no longer politically correct in this current age were looked upon as being less offensive and more socially acceptable back then.

I credit a lot of things that happened back then to the aviator mentality. In the beginning God created the aviator, and the aviator turned to God and say, "Thanks, I've got it; I'll take it from here." And God became the copilot and the aviator made the rest of the world and the rules by which we all lived. Most of those serving in the ranks knew the phrase, "It's a pilot's Air Force."

I think the retired stag bar club managers must have transitioned into the public relations jobs in Las Vegas. You know? They must be the ones that came up with the slogan "What happens in Vegas, stays in Vegas." The saying has often been applied to stag bars.

When I joined the Air Force in 1968, stag bars were common on most bases and SAC bases were no exception. While some may have just been backroom bars where fliers could go get a brew after a long flight, some were much, much more than that. I was hampered by the handicap that I did not drink, but I did enjoy socializing with those that did, so I was not totally naive. There were plenty of bar snack foods such as popcorn and peanuts available, and hamburgers, hot dogs, and French fries were on the menu for guests who needed more substance.

My first thoughts when I try to remember things from my second lieutenant days of the late Sixties were some of the very strict rules that were usually learned through the "combat rules" of the establishment. The first rule that comes to mind helped give the bar the moniker it acquired. It was the "stag bar." The number three dictionary definition of stag is "a man who attends a social gathering unaccompanied by a woman." We're not talking about male deer, but unaccompanied male fliers. No women were allowed in the stag bar; it was off limits to them – even the female officers of the time could not enter. This was long before the age of people being worried about political correctness and sexual equality. The stag bar was for males. Period. Heaven help the crewdog who had a wife who came looking for him and crossed the hallowed doorstep to this male dominated kingdom. Not only would the missing husband have to answer to his wife, worse yet, he would be the recipient of the wrath of his fellow fliers, and he would also suffer the financial humility of having to buy a round of drinks for all that witnessed the act.

The next rule that threatened the wallet of a deviator of the rules was characterized by a large brass bell that hung over the bar. If someone became the reason for the bell to toll, then that person had to "buy the bar" as well. Some victims tried to appease the Gods by purchasing a couple of pitchers of beer to be passed around, but others would find that it took a greater sacrifice of monetary funds to pay his debt to society. Individual drinks were often the penalty for a transgression.

And what crime was the greatest cause for the bell to be rung? Coming into the stag bar wearing your "cover" – your hat! If you failed to remove your hat before you walked in, or put it on your head before you walked outside, the wrath of the bell would fall upon you. People could not say they were not warned. Hanging on the wall in plain sight was a large sign proclaiming "He who enters covered here, buys the bar a round of cheer!"

I remember standing in the stag bar at Mather AFB one day and an officer wearing his flight cap walked in and headed straight toward the bell. Before the bartender or any of the other patrons could get to the bell, he reached up and rang it himself. It seems that he had just found out he was on the promotion list for major, and that was his way of celebrating.

Fighter aircraft bases were the most memorable, but Mather's stag bar was the most notorious one I ever visited on a base that had a SAC component. Now the crews from SAC cannot be blamed for all the activities that took place there, because Mather was also the home of the Navigator Training Wing and thousands of young second lieutenants passed through its gates annually. Many of those young fliers were fresh out of college and drawing a real paycheck for the first time in their lives. A housing area known as "Sin City" was even habituated by groups of bachelor nav students sharing rooms instead of having individual BOQ rooms. When I was there it was California and it was the late Sixties. Its location was only a couple of hours to the east of San Francisco and Haight-Ashbury and hippies and the craziness that defined the social "free-love" and "burn your bra" freedom was approaching its sunset, but was still glowing brightly. San Francisco night clubs were the spearhead of the phenomenon of topless waitresses and dancers.

The Mather stag bar was located in the rear of the Officer's Club. In front of the club there was a parking place with a sign stating it was "Reserved for the Wing Commander or any Second Lieutenant." In the rear of the club, all officers, regardless of rank were equals, except female officers of course. I never knew if it was a grand scheme or just the luck of the roll of the dice, but two contrasting activities were scheduled for Wednesday nights at the club. In the main club, Wednesday night was bingo night. Many of the officers' wives enjoyed the games of chance that offered large cash rewards, and many would drag their husbands to the club with them on bingo night. In the meantime, unknown to most wives, another activity was taking place back in the stag bar of the same club. As I stated, it was California and it was the Sixties. It was a time of the sexual revolution and it was the era of topless waitresses and topless go-go dancers. There was even a barber shop in Sacramento that featured topless female barbers. Topless and more was on the bill for the stag bar.

I remember a gentleman who normally wore a light colored suit and walked with the aid of a cane who would supply the girls for Wednesday night's activities. Members of my navigator training class jokingly called him "the pimp" and for all we knew he could have been one. Anyway, he would escort the lovely ladies to the bar and sit in the back while they performed their routines on a small stage in the corner of the bar, then escort them out at the end of the night. The stage really was very small, and the crowds were very large on most nights. When the girls performed, the crowd stood around the stage and did not say seated in the few small tables available. This was before the invention lap dances so the rule was no physical contact, unlike some of the bars in Newland near U-Tapao in Thailand. Sometimes there was one dancer and sometimes two shared the stage. I remember one night a girl dressed in street clothes sat back with the pimp while the featured act did her routine on stage. The girls usually wore costumes, and though they were not strippers in the vaudeville sense, the clothing covering their bodies was shed in time until they were dancing totally nude on the stage. Yes, not just topless, but totally nude. That evening the visitor got inspired by the activities and eventually she also took the stage and though not prepared in costume, stripped bare naked, while the rock and roll music blared from the speakers and the crowd cheered her on. In the meantime, the wives in the main club continued to enjoy their bingo game, naively unaware of what was really happing when their husbands disappeared saying they were going to the bar for a drink.

Wednesday was not the only night when the nude dancers performed in the bar. They also were the scheduled act on a weekend night. My first thought was Friday night, but then I remember that on Friday night the "happy hour" activity in the main club was the main attraction. The nav students and their dates and wives would meet there and enjoy cheap drinks and dance to the music of local bands. One band that stands out in my memory was comprised of four very large females who called themselves "The Jolly Girl Giants." So, it must have been Saturday nights that the ladies bared all to entertain the troops in the stag bar. USO shows were never like that.

When I graduated from EW school in 1971 and headed for Castle AFB down the road, the Mather stag bar was still going strong. I do not know when stag bars finally fell to the constant pressures of the wives' club or other outside sources who failed to recognize or appreciate the entertainment value they presented to the officer corps. Eventually the stag bar faded into history.

One final note about my memories of the Mather stag bar comes to mind. When I went to Guam with Operation Bullet Shot, I was sitting in the officer's club eating lunch, accompanied by a couple of lieutenant colonels I worked with in bomber ops. We looked up and a young captain came walking in with a pretty blond in a revealing island dress that accented her curves accompanying him. One of the colonels remarked about how gorgeous she was and I took a closer look. You should have seen the look on their faces when I calmly stated, "You think she looks pretty like that? You should see her without the dress in the nude." She had been one of the topless dancers at the Mather stag bar and someone told me that she had married an officer going through electronic warfare training. And they say it's a pilot's Air Force. Yeah, sure.

Faces

George Thatcher © 2011

Tonight we note, with deep regret,
The absence of a cherished few,
Whose faces won't be seen again,
Till next we meet in life renewed.

Their faces seem to linger still,
Inscribed upon our memory,
And, what a book their lives would fill!
Each page unique in history.

It's said they were a handsome lot,
Whose profiles shone so admirably,
But they, who wore them, gave it not
a thought; engaged were they, as we,

with living ordinary lives,
In those extraordinary times
of violence and fragile peace,
While tyrants plotted heinous crimes

against the lost and long-deprived:
the innocents. And they were tasked
to speed relief, to stem the tide,
And serve the cause wherever asked.
Their scant reward: ingratitude,
Most notably, from their own side.

Too soon they've gone, with little sound
To mark their passing, save the winds
of constant change, that sear the ground,
Erasing deeds, remitting sins.

For us, still here, this much remains,
Which wind or wave cannot erase,
Which time preserves, in hearts engrained:
The memory of each noble face.
And now they re-emerge to share
An evening of camaraderie,
A drink, a laugh, with friends of old,
Before they fade to history.

And we'll reprise this reverie,
As long as one of us draws breath,
Until we join the gallery
of faces, long alive in death.

Always a Crewdog

George Thatcher © 2011

The times we shared in olive drab,
Strapped in the traces like old sled dogs,
The moments of stark terror
Interrupting our long hours of blissful routine.
The screeching of the klaxon,
Calling us to crank 'em up
and wait for the dreaded Message
(which, thankfully, never came).

Long weeks in the alert facility,
Where we honed our skills daily
at poker and other mandatory drills, while
bravely abating our barley-based thirst
until the changing of the guard.

Where we played the career game,
And sweated out the promotion list,
Knowing that some would suffer
the ultimate humiliation, might even
be forced out of the Corps and be
banished back into the ranks.

Was any of this fun?
Did we curse the Commander,
The incompetent staff weenies,
The hangar queens that could
Never put a bomb in the box?
Did we dread the next assignment
to someplace even worse than
this godforsaken hole?

Did we dream of conquering vast new fields
as civilians? Would the world
rejoice at our arrival into
the hallowed halls of the private sector?
Did we set production records,
make millions, soar to heights unseen,
or just sink into three-piece
anonymity?

And did the corporate gods
take note of our rows
of ribbons from the glory days?
Maybe once a year, on Veterans Day.
Maybe on those biennial occasions
when we gathered at reunions,
To tell the old stories over and
over again, until our wives
knew them all by heart, and were able
to reign us in when we
engaged in retrospective hyperbole.

Yeah, looking back, we weren't really
all that invincible: neither head of
the class nor Saviors
of the Free World.
But we did our jobs, gave our youth,
paid the price.
Not heroes, just regular guys who
ground out the years
in a combat crew slot.

Would we change it now? Hindsight
bestows vision we didn't have then,
Shows us forks in the road that were once
invisible, illuminates choices with
crystal clarity, points to what we
shoulda, coulda, mighta.

And at the end of that long day,
What we're most proud of,
While yet unsure as to whether
or not we'd choose to repeat
it all, is having been
Crewdogs.
That remains the constant
in our lives. The rest is all
written in lower case.

And if there's still any doubt
as to who we are, it just
boils down to this:
Once a Crewdog,
Always a Crewdog,

And don't by God forget it!

Résumé

Résumé [rez-oo-mey] - *noun* - brief written account of personal, educational, and professional qualifications and experience.

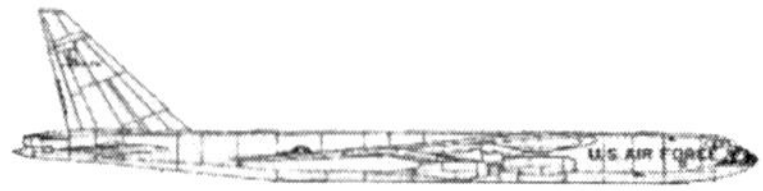

The Contributing Authors

John R. Bair

Former Captain, USAF, Entered the Air Force through OTS, Lackland AFB, TX, 5 Nov 68. I completed UNT Navigator Training at Mather, AFB CA, in Sep '69. I then went to Electronic Warfare Training graduating in Apr '70 and afterward completed AF Survival training, Fairchild AFB, WA, May '70, followed by CCTS in the F-model, Castle AFB, CA, Aug '70.

I was assigned to B-52Hs at Kincheloe AFB, MI, 716BS, 449 Bomb Wing in Sep '70 to crew E07. My first TDY was to B-52D Arc Light RTU, May '71 and then proceeded to U-Tapao, Thailand, passing through Guam, assigned to 307 Strategic Wing, May – Sep '71.

I separated from active duty at the end of my active duty commitment 15 Sep '73.

As a side note, I would have had enough time for a second Arc Light tour in late '72, but because of the closeness of my DOS after return and a volunteer EW for my slot, Capt Nutter Wimbrow, I was spared. Capt Wimbrow, a close friend, was shot down during Linebacker II, KIA, 26 Dec '72.

Clif Banner

Lt Col, USAF (Ret.) earned his Bachelor's from University of Massachusetts and his Masters at University of Nebraska. Drafted by lottery in his senior year, he earned his commission through OTS. He completed Undergraduate Pilot Training at Craig AFB and B-52 pilot certification at Castle AFB in 1972. At his initial operational assignment in the B-52D, Westover AFB, he deployed to Andersen AFB and U-Tapao RTAB under Operations Arc Light and Bullet Shot. When Westover closed at the end of Vietnam, assigned to Griffiss AFB, he returned to Guam and Thailand for post-Vietnam contingency flight operations and alert duty.

An instructor and evaluator pilot at Griffiss, he was reassigned to Barksdale AFB as MAJCOM evaluator at 1st Combat Evaluation Group from 1978-1981, flying the B-52D, B-52G, and B-52H. At Headquarters SAC at Offutt, Clif was responsible for developing SAC-wide alert generation procedures; he flew the T-39A for several years, culminating 13 years of flying duty with over 3,000 hours. As the command nuclear weapons officer, he scheduled and tracked SAC's weapons inventory, and, in 1984, became JSTPS manager for targeting, terrain, threats, and other data required for SIOP planning. He led the mission planning efforts during early development of the B-2 and was manager for the first automatic routing systems for cruise missiles. At the Joint Staff (JCS), Pentagon, from 1988-1991, he was responsible for three nuclear operations plans and was Lead for both US and SHAPE Nuclear Planning Systems. He returned to Offutt as Chief, Mission Planning Division, and he helped stand up of the US Strategic Command when SAC was disestablished; he retired from active duty in 1993. Clif has continued his contributions to the Air Force as Chief Engineer, Joint Expeditionary Force Experiment at Electronic Systems Center, Hanscom AFB. He lives in Nashua, NH, with his wife Linda Laine Banner and two sons, Benjamin and Aleks.

James E. Bradley

Lt Colonel, USAF (Ret.) graduated from college in 1959, with a Bachelor of Science in Mathematics and Chemistry. He enlisted in the United States Air Force, completing basic training in October of 1959 at Lackland AFB, Texas. He was assigned to the Air Force Special Weapons Center, Physics Division's Pulsed Power Laboratory at Kirtland AFB, New Mexico. There he performed Engineering and Scientific Aide duties. He was promoted to Airman 2nd Class during this assignment.

He entered Officer Training School (OTS) in April of 1961 at Lackland AFB, Texas, and was commissioned a 2nd lieutenant on 27 June 1961. He was assigned to Undergraduate Navigator Training at James Connelly AFB, Texas, training on the venerable T-29 Flying Classroom. He received his navigator's wings in April 1962.

He entered Electronic Warfare Officer Training at Mather AFB, Sacramento, CA, and graduated in February 1963. He completed USAF Arctic Survival Training at Stead AFB, Reno, Nevada. He completed B-52 Combat Crew Training at Castle AFB, Merced, California, in July 1963, and was assigned as an EWO on the B-52D at Glasgow AFB, Montana.

He served with the 91st Bomb Wing, 322nd Bomb Squadron until the unit was disbanded in spring of 1968. Lt Col Bradley has approximately 3,400 hours total flying time with 2,356 hours in the B-52B/C/D/F.

Lt Col Bradley completed the USAF Iron Hand/Wild Weasel Training in August 1968. He was assigned to the 355th Tactical Fighter Wing, 333rd Tactical Fighter Squadron at Takhli RTAFB, Thailand, as F-105F EWO/"Bear." Upon return he entered the Air Force Institute of Technology sponsored Graduate School at Texas A&M University to study Computer Science. Graduating in December 1970, he was assigned as Computer Operations Branch Chief, Headquarters Air Training Command's Data Automation Division.

He completed Armed Forces Staff College in June 1974 and was assigned to the U. S. Readiness Command/J5, as a Joint Operations Planning System Officer, MacDill AFB, Tampa, Florida. Next he was assigned to Headquarters, Tactical Air Command (TAC) on the Planning Staff of the DCS/Plans.

After retirement in February 1980, he worked for the Boeing Military Airplane Company, Wichita, Kansas, on the B-52 Modernization Program as a Software Test Engineer. He tested the Offensive Avionics System's software in the laboratory prior to flight in the B-52G/H. He also tested flight simulators and mission planning software. He retired from Boeing in 1995. Jim and his wife, Marian, live near Westmoreland, Kansas.

Kenneth G. Brown

Major, USAF (Ret.) Entered the Air Force via the Aviation Cadet program in March of 1957. Completed navigator training at Harlingen AFB earning both a commission and flight wings. Advanced training at Mather AFB followed by flight training in the B-47 at Little Rock AFB and PCS to Hunter AFB's 20th Bomb Squadron (2nd Bomb Wing). Four years in the B-47 earned him a regular commission and spot promotion. Phase out of the B-47 in 1963 resulted in reassignment to Glasgow AFB's 322 Bomb Squadron flying B-52Ds. Following the 1966-67 Arc Light tour Ken PCS'd to Colorado State University under the AFIT program and finished his college degree.

After graduation and an incomplete preparation for an EB-66 assignment Ken spent a year in Southeast Asia (NKP) flying "Candlestick" in the C-123. Back in the states he spent a year in NORAD's Cheyenne Mountain and three years in Denver in a computer operations position before being reassigned to Robins AFB (19th Bomb Wing). Ken retired in October of 1977. He can be reached via e-mail at: kennlee@windstream.net

William J. Bushman,

Lt Colonel, USAF (Ret.) After high school I enrolled at Iowa State College and joined the AFROTC. Later I transferred to the State University of Iowa. Here I completed my undergraduate studies, received my BA and my commission.

My first assignment was to Lackland AFB for pre-flight indoctrination training and physical prior to proceeding to my UPT base of Bainbridge, Georgia. This was followed by BPT at Craig AFB, Alabama. My choice of aircraft after UPT was the B-52.

After completing all prerequisite training, I reported to Bergstrom AFB, Texas where I spent the next 5 ½ years. I served as copilot, Stan/Eval copilot, and aircraft commander during this time.

From March AFB, I was deployed four times to Southeast Asia for Arc Light missions, flying 150 of these out of Guam, Okinawa or Thailand. Shortly after upgrading to Instructor Pilot I was found to have cataracts developing and flew for two years on a medical waiver. In 1971, I went to Maxwell AFB, Alabama to attend ACSC. After graduation from ACSC and getting my MBA from Auburn University I was forced into accepting an operation on my eye and subsequently grounded as a result.

I chose aircraft Maintenance and went to Luke AFB, Arizona where I was Maintenance Supervisor, Chief of QC and Squadron Commander. Later I transferred to Kunsan AB, Korea as a Maintenance Squadron Commander and ended up as Base Commander.

Moody AFB, Georgia was next. There I was Assistant DCM and Deputy BC. My final active duty assignment was to 14th AD at Beale AFB, California as Deputy DL.

My days as a "crewdog" were undoubtedly the happiest ones I spent on active duty and the memories of them and the wonderful crew members I flew with will remain with me forever.

Gene Cassanova

Major, USAF (Ret.) was born July 2, 1947 in New Orleans, Louisana. In 1965, after graduating from Metairie's East Jefferson High he attended Louisiana State University where in 1970 he earned a B.S. degree in Psychology. He earned two Masters Degrees: an MLA from TCU in 1977 and another in Industrial Technology in 1981 from East Texas State University. He graduated from Squadron Officers School, Air Command and Staff College, Air War College, and National Security Management.

Major Cassanova received his commission through the Officer Training School, Lackland AFB, Texas, in 1970. He proceeded to Reese AFB and completed Pre-Undergraduate Navigator Training. He earned his navigator wings at Mather AFB, California, in September 1971. Next he attended Navigator Bombardier Training at Mather and was awarded the Lenzi Trophy for excellence in simulator training. Following training in the B-52 at Castle AFB, Major Cassanova was assigned to the 367th Bombardment Squadron, McCoy AFB, Florida. Six months later he received PCS orders to Carswell AFB, Texas, and was assigned to the 9th Bombardment Squadron, 7th Bomb Wing in positions from B-52D and B-52H Navigator, through Instructor and Evaluator Radar Navigator.

In 1982, Major Cassanova was assigned to the 1st Combat Evaluation Group (1 CEVG), Barksdale AFB, Lousiana, as Chief, RBS/EW Equipment Development. During his tenure in 1 CEVG and later the 1st Electronic Combat Range Group (1 ECRG), he directed a 15-member acquisition staff in the development, testing, evaluation and procurement of 38 electronic combat range systems valued in excess of $600,000,000. He was instrumental in the largest deployment of aircrew training systems for the Strategic Air Command. He has flown in the B-52D, F, G and H-models during his career.

He is married to the former Marcia Anne Wehmeier and has one daughter Elisabeth Reagan.

After his second 20-year tour of duty, this time with the United States Attorney's Office, Gene is now retired... again.

Allan F. Cobb

Lt Colonel, USAF (Ret.), grew up in New Carrollton, Maryland with some time at Griffiss AFB, New York. His father, TSgt Keith V. Cobb, Ret. was assigned to the 416th Bomb Wing command post in 1966-67. He was commissioned through AFROTC at the University of Maryland, College Park, in 1981, graduating with a BS in Computer Science.

His first assignment was at Electronic Systems Division, Hanscom AFB, Massachusetts. He became a private pilot, then competed for Undergraduate Pilot Training. After UPT, he was assigned to the B-52H, 20th Bomb Squadron, Carswell AFB, Texas, in February of 1985.

He upgraded to aircraft commander in 1987, and to instructor in 1988. He was assigned to the squadron training flight, then as a Stan/Eval flight examiner.

In early 1990 he was assigned to the 328th BS, Castle AFB, California, as a B-52G CCTS instructor pilot. He was temporarily assigned to the 1500th Provisional Bomb Wing, Andersen AFB, Guam, supporting Operation DESERT STORM and the 4300th Provisional Bomb Wing, Diego Garcia.

In 1992, Lt Col Cobb was accepted to fly the U-2 high-altitude reconnaissance aircraft at Beale AFB, California. His remaining months at Castle were as the assistant pilot branch chief.

While flying the U-2, he was assigned to the 99th Reconnaissance Squadron, and its operating locations, including positions as 9th Operations Support Squadron operations officer, HQ Pacific Air Forces, Hickam AFB, Hawaii, and in 1999 as Commander, 363rd Expeditionary Reconnaissance Squadron, Prince Sultan AB, Saudi Arabia. Then he became chief of the Korean Reconnaissance Center (a U-2 flight position), HQ, 7th Air Force, Osan Air Base, Republic of Korea.

Returning again to Beale AFB, he retired in June, 2004 with 4900 flying hours, including over 2700 in the B-52G/H and 1650 in the U-2R/S.

Lt Col Cobb resides in Rockville, Maryland, married to Hsiaoming Yao, and flies for Southwest Airlines. He has two daughters, Katherine and Elizabeth and a son, Jason.

Doug Cooper

Lt Colonel, USAF (Ret.), earned a B.A. degree in 1965 at the University of Utah. Enlisted in the U.S. Air Force after graduation when he was a first draft pick for the U.S. Army. Earned a commission at Officer Training School in February of 1966. Attended Navigator Training and Electronic Warfare Officer Training at Mather AFB, California. Following B-52 CCTS training at Castle AFB, California, he was assigned to the 744th Bomb Squadron at Beale AFB on a B-52G crew and then served for six months on a B-52D replacement crew in bombing missions from Guam, Thailand and Okinawa.

Things went from bad to worse and he was PCS'd to Carswell AFB just in time for Bullet Shot where he participated in four more deployments. He logged over 3,000 hours in B-52s and over 5,000 hours total flight time. In 1974, he was offered a job as a unit command post officer controller in the Carswell Command Post. After his promotion to Major, he was assigned to the Strategic Air Command Airborne Command Post (Looking Glass) for three years where he was promoted to Lt Col, followed by assignments to Zaragosa AFB, Spain, Incirlik AFB, Turkey, and finally to Mather AFB. He held the position of DOC in these last assignments.

His decorations include the Meritorious Service Medal, the Air Medal with six oak leaf clusters, the Armed Forces Expeditionary Service Medal, the Republic of Vietnam Campaign Medal and the South Vietnam Cross of Gallantry with Palm. After retiring from USAF, he worked as a program consultant for the California Department of Education performing audits of California public school districts. He and his wife, Susan, live in Lincoln, California, in a Del Webb community. He is active in veteran organizations, travels extensively and flies a 1969 Cessna 150 around the Sacramento Valley when the mood strikes him.

Gene Daspit

Lt Colonel, USAF (Ret.) earned his B.S. degree in Civil Engineering in 1968 from the University of Wisconsin Madison and was commissioned a 2nd lieutenant through AFROTC. He attended Navigator Training and Navigator Bombardier Training at Mather AFB, California, then completed B-52 Crew Training at Castle AFB, California flying the B-52F. In 1970, he was assigned as a crew navigator to the 46th Bomb Squadron, Grand Forks, North Dakota flying the B-52H.

In 1971, his crew deployed to U-Tapao RTNB, Thailand flying 45 Arc Light missions in the B-52D. In December 1972, his crew again deployed flying 55 Arc Light missions from U-Tapao and Andersen AFB, Guam. During this period, he flew six Linebacker II missions from U-Tapao.

Upon return to Grand Forks, he upgraded to radar navigator, serving as instructor radar navigator on a crew and in Training Flight. In February 1976, he was assigned to the 60th Bomb Squadron, Andersen AFB, Guam flying B-52Ds as the radar navigator on the senior Stan/Eval crew, then becoming the 43rd Strategic Wing Target Study Officer. In 1979, he was assigned to the Data Automation Computer Facilities Branch at Headquarters SAC. In 1983, he was assigned to the 325th Bomb Squadron, 92nd Bomb Wing, Fairchild AFB, Washington, as a crew instructor radar navigator and then Wing Target Study Officer.

On October 16th, 1984, while flying as substitute radar navigator, he survived a low-level training accident in Monument Valley, Arizona. While recovering from injuries sustained in the accident, he was assigned to Operation Plans until his retirement in 1991.

He logged over 4,000 hours in the B-52 D, F, G and H-models. His decorations include the Distinguished Flying Cross with oak leaf cluster, the Meritorious Service Medal with two oak leaf clusters and the Air Medal with five oak leaf clusters. After retirement, he worked as a Civil Engineer and now is fully retired, devoting his time to training his two Golden Retrievers to compete in American Kennel Club events in obedience, tracking and agility.

Derek H. "Detch" Detjen

Major, USAF (Ret.) grew up in New York City and in the Northern Kentucky/Cincinnati area. As an aeronautical engineering student at the University of Cincinnati, he flew in one of the first KC-135As during its cold weather testing in Alaska. Entering the Aviation Cadet program at Harlingen, Texas, in late 1960, he finished as a Distinguished Graduate. EWO school at Keesler AFB, MS followed where he met his wife Betty.

Assigned to Turner AFB, Georgia, he participated in the first six-month Arc Light tour to Guam in 1966. Returning in both 1967 and 1968 as a crewmember and staff EWO briefer from Columbus AFB, Mississippi, he was part of the first crew to complete 100 Arc Light missions in November of 1967. Crew E-13's 5th Air Medals were presented to them in Guam by the then CINCSAC, General Paul McConnell.

Assigned to the Castle AFB, California Replacement Training Unit (RTU) in July of 1969, he spent a rewarding four-year tour, training the G and H-model crews on Arc Light tactics/procedures. Later while in 1CEVG, another five-month tour on Guam ensued at Det. 24 "Milky," training the recently deployed B-52 wing. His last five years were spent at Barksdale AFB, Louisiana, in charge of B-52 and KC-135 EWO study, including a trip to RAF Fairford, England during the 1982 Crested Eagle NATO exercise. His military decorations include a Distinguished Flying Cross, eight Air Medals, a Meritorious Service Medal, two Combat Readiness Medals, an Air Force Commendation Medal and several lesser awards.

Major Detjen worked for GD on the Trident submarine at NSB Kings Bay, Georgia, for five years, attained a graduate degree from Valdosta State University and taught in their on-base education program. A final nine-year stint at Aiken Technical College in South Carolina saw him running their Management and Marketing majors before his retirement in 2000. A lifelong devotee of The Masters, he now lives in Evans, Georgia, about 10 minutes from the Tournament, which he has managed to attend for over 40 years.

Richard L. Gaines

Colonel, USAF (Ret.) Col. Gaines retired in 1983 after 27 years of service, all but three of which were in SAC. He piloted B-52D, B-52E, B-52F, B-52G, and B-52H aircraft for over 6,000 hours and the F-86 and B-47 for another thousand. He, his wife Bunny and children Kristi and Lee were stationed at McCoy AFB, Florida, Walker AFB, New Mexico, Grand Forks AFB, North Dakota, Castle AFB, California, Seymour-Johnson AFB, North Carolina, Loring AFB, Maine, and Dyess AFB, Texas. Over those years, Colonel Gaines held most squadron and wing positions including two years as OMS Commander at Seymour-Johnson AFB . He flew more than 50 combat missions as pilot in the B-52D during the Vietnam conflict. His alert pulling crew duty extended from 1959 to 1976. Among other honors, his crews won the Eisenhart Trophy for the best B-52 crew in 15th Air force and another was selected to represent Grand Forks in the 1970 Bombing and Navigation Competition at McCoy AFB, Florida. From 1980 to 1983 he was assigned as Director of Operations for the 68Th Bomb Wing at Seymour-Johnson, Director of Operations at the 42nd Bomb Wing at Loring and held the same position at the 12th Air Division at Dyess.

Russell Greer

MSgt, USAF (Ret.) Born in Jacksonville, North Carolina, in 1959 and moved to Tuscaloosa, Alabama, in 1969 when my father was transferred there with Sears. Enlisted in 1977 serving as an Automated Tracking Radar Systems Technician stationed at Det. 10, 1CEVG, Hastings, Nebraska.

Cross-trained in 1982 to B-52 gunner. CCTS instructors were Anthony Freeborn and Bruce Helyer; flight line instructor was Mike Riggs. Assigned to the 62nd BS, 2nd BW, Barksdale AFB, Louisiana, Crew R-06. I attended CFIC in 1985 and in 1986 I was assigned to S-03 for the 2nd BW. In 1987 became S-01 gunner. Participated in 1985 and 1986 SAC Bomb-Nav Competition where our crew finished third in the bombing portion even though we were the only crew still equipped with the ASQ-38. In 1988 I was selected by Headquarters SAC to relocate to the 436STS, Carswell AFB, Texas, to become the B-52G Curriculum Development Manager. Completed H-model qualification and maintained G qualification with monthly TDYs to G units. Deployed with the 806th BW (Provisional) to RAF Fairford during Desert Storm as a Staff Planner and gunner. Flew numerous combat sorties as lead gunner. B-52G/H, Instructor and Evaluator.

When HQ SAC removed the gunners from the B-52 in October 1991 I retrained into Flight Engineer, attended FE training at Altus AFB, Oklahoma, and was assigned to C-141s at McGuire AFB, New Jersey.

Took an early retirement in December 1993 after the death of my parents in an aircraft accident to take over the family business in Alabama. Raced the ARCA Superspeedway Series from 1995 to 1997 competing at tracks such as Talladega, Charlotte, Michigan, Daytona and Pocono. Instrument Rated Pilot with over 3,000 hours in B-52s, 1,000 in C-141s and over 800 as a private pilot. A very proud father of three boys and two girls, Brian, Michael, Emory, Aimee and Emilee. I am married to Dianne, the light of my life. Currently reside in Lincolnton, North Carolina. Franchise Consultant for Aaron's Sales and Lease. ROLL TIDE ROLL!

Jack C. Hawley

Major, USAF (Ret.) was born and raised in the western suburbs of Chicago. My early teen years surfaced an interest in music and an opportunity to fly in a Cessna 182. That set the aviation "hook". Following high school graduation, I was accepted at Southern Illinois University (SIU), to pursue flight through AFROTC and a BS in Marketing.

Freshman year, I "lucked" into a senior ROTC trip to Eglin AFB; setting the hook a bit deeper. The SIU Sport Parachute Club allowed me to jump out of perfectly good airplanes. I applied for USAF pilot training while in Advanced ROTC and with 23 other cadets, completed the Flight Instruction Program, prior to graduation and commissioning. Four of us earned our private pilot's license.

Undergraduate Pilot Training occurred at Reese AFB, where I graduated in absentia. Arlene and I were married that 28 Oct 67, 1,000 miles away. The standard training package for those BUFF-bound, prepped me for initial certification as a copilot at Carswell in June of '68.

Four months later our crew was westbound, headed for the Vietnam Arc Light theatre of operation. I participated in six tours (three as copilot, three as aircraft commander) from 1968 through 1973 inclusive. I racked up 380+ combat sorties and 2,100+ combat flight hours. Luckily, I was home, when our first son showed up in Sep '71.

Post-Vietnam assignments included Squadron Officer School, Command Post duty at Carswell and Goose Bay, Labrador, (where our second son was born in May of '75). I joined the "rated supplement" as OMS Supervisor at Kincheloe and Seymour Johnson, where I returned to the cockpit as crewdog IP. With SAC contingent closure at Seymour Johnson, I returned to Carswell as a Squadron IP, also working out of Training Flight, the simulator, and sole member of the Instrument School, retiring 31 Oct '87.

Post-military jobs were basically aviation technical writing related; also writing for Nokia, managing a pharmaceutical laboratory, and an 18-wheel semi, coast-to-coast, driver.

Residing west of Fort Worth for the past 26 years+, married for 41 years+, we're "gramma" and "gramps" to four terrific grandchildren.

Stephen "Mulligan" Henley

Former Captain, USAF, was commissioned in 2003 through Texas A&M University's Reserve Officer Training Corps Program. He has active duty flight time in the MH-53, T-1, T-37, T-43, and B-52H. He was initially assigned to Hurlburt Field, Florida, as a Communications Officer, where he served two tours in the Middle East at the Combined/Joint Special Operations Air Component Headquarters as the J6/Deputy J6 and the Executive Officer.

After completing an assignment as the 16th Operations Group Executive Officer, he was selected to the Euro-NATO Joint Jet Pilot Training (ENJJPT) program at Sheppard Air Force Base, Texas. Realizing the unlimited potential of this young aviator, the Air Force reassigned him to Undergraduate Navigator Training (UNT) at Randolph Air Force Base in 2007. After completing UNT with a specialty in Electronic Warfare in 2008, he graduated as a Combat Systems Officer (CSO). He was assigned to the 11th Bomb Squadron at Barksdale AFB where he graduated the FTU (Formal Training Unit) for B-52Hs in July 2009. Captain Henley was assigned to the 23rd Bomb Squadron at Minot AFB where he served as the unit's Deployment Manager for their deployment to Guam. He also served as the squadron's Weapons & Tactics EWO and a Flight Commander. In 2011, the AF RIF board awarded him a promotion three BTZ to civilian and he separated from the Air Force in March 2012. He, his wife Christine, and his two sons (Jonathan and Jackson) currently reside in New Braunfels, Texas, and are exploring future career options.

Thomas D. Jones, PhD

Former Captain, USAF, is a veteran NASA astronaut, scientist, speaker, author, and consultant. He holds a PhD in planetary sciences, and in more than 11 years with NASA, flew on four space shuttle missions to Earth orbit. On his last flight, Dr. Jones led three spacewalks to install the centerpiece of the International Space Station, the American Destiny laboratory. He has spent 53 days working and living in space.

Tom is a Distinguished Graduate of the U.S. Air Force Academy. He piloted B-52D strategic bombers, studied asteroids for NASA, engineered intelligence-gathering systems for the CIA, and helped develop advanced mission concepts to explore the solar system prior to joining NASA's astronaut corps.

Tom's latest title is *Planetology: Unlocking the Secrets of the Solar System* (written with Ellen Stofan, PhD; National Geographic, 2008). His other 2008 book, *Hell Hawks!* (with Robert F. Dorr; Zenith Press), is a true story of an aerial band of brothers in WWII, and the top-selling title at the National Air & Space Museum. *The Wall Street Journal* named his *Sky Walking: An Astronaut's Memoir* (Smithsonian-Collins, 2006) as one of its "Five Best" books about space.

Dr. Jones' awards include the NASA Distinguished Service Medal, four NASA Space Flight Medals, the NASA Outstanding Leadership Medal, the NASA Exceptional Service Award, the NASA Exceptional Public Service Medal, Phi Beta Kappa, and the Air Force Commendation Medal.

Tom served on the NASA Advisory Council and is a board member of the Association of Space Explorers. As a senior research scientist at the Florida Institute for Human and Machine Cognition, he is currently active in asteroid exploration, planetary defense from asteroids, and in the debate over our nation's space exploration policy. His website is www.AstronautTomJones.com

Kenneth P. Katz

Former Captain, USAF, earned a degree in aeronautics and astronautics from the Massachusetts Institute of Technology in 1985 and was commissioned as an officer through the Air Force ROTC program. In October 1985, he reported to the Air Force Flight Test Center, Edwards AFB, California, and spent the next four years assigned to the 6520th Test Group and 6519th Test Squadron as a flight test engineer, supporting B-52 and missile flight test programs.

In 1989, he was the instrumentation operator during the first launch of an AGM-136A Tacit Rainbow missile from the B-52. After separating from active duty service, Ken worked for Boeing as a flight test engineer on the V-22 Osprey tiltrotor program and then attended graduate school at the University of Michigan to earn graduate degrees in aerospace engineering and manufacturing.

He is currently employed as a project manager in the information technology and software development fields. Ken currently lives in Longmeadow, Massachusetts with his wife of 20 years and their three children. He has a commercial pilot certificate with instrument rating and 1,000 pilot hours, and flight experience as a flight test engineer and observer in over twenty types of military aircraft. In 2007, his book *B-52G/H Stratofortress in Action* was published, with a second edition published in 2012.

Harold "Rusty" Keller

MSgt, USAF (Ret.) Born in Torrance, California. Raised on a working cattle ranch near Bakersfield, California. Served five years in USMC as an aircraft hydraulic systems mechanic on F-8, F-4, and CH-46 aircraft. Flew 214 combat missions in Vietnam as a CH-46 helicopter gunner with the Marine Corps. Served 15 years with the USAF, which included: two years as an aircraft crew chief on KC-135A, and B-52D aircraft. Nine years as a B-52 Gunner (B-52F Castle, B-52G Fairchild, B-52H Grand Forks, and B-52D Andersen). Five years as a C-130 Flight Engineer, (C-130E Little Rock, C-130H Dyess). USAF Vietnam support experience includes: one Young Tiger (U-Tapao), and two Bullet Shot (Andersen) deployments, which also involved Linebacker II. All of these were aircraft maintenance support activities. After retirement, and during Desert Shield, worked with B-52G WST at Andersen (Pilot's and Defensive team consoles).

Since then, work areas were: general aviation commercial pilot, commercial aviation jet mechanic, some cowboy type jobs, and about 30 magazine articles published. Married, with three children, and six grand-children. Member of: The Marine Corps League, American Legion, American Legion Riders, and Viet Nam Veterans of America. Current home is in New Mexico.

Roland R. LaFrance, Sr.

Former SSGT, USAF Born July 22, 1954, Nashua, New Hampshire. My father was a City firefighter. Graduated Bishop Guertin H.S., June, 1972. Inducted into the U.S. Air Force under delayed-enlistment program, April, 1972. USAF Basic Training, June to August, 1972, was honor graduate. Attended USAF Fire Protection Specialist School (57130) at Chanute AFB, IL. Graduating, October, '72.

Assigned to 42ND CES Fire Dept., Loring AFB, Maine. Rapidly completed training to become a 57150, Fire Protection Specialist. July, 1973: TDY assignment under "Bullet Shot" to Andersen AFB, Guam, housed in barracks area seven miles from base. Returned to Loring AFB after October Mid-East war shortened TDY.

1974: Met my future wife, Francesanne Loveless, daughter of an assistant fire chief. Promoted to E-4 (Sgt). Married in Feb., '75 at base chapel. As secondary rescue crew chief, became an Emergency Medical Technician-Ambulance.

Left active duty in June, 1976. Assigned to active Reserve firefighting flight, Westover AFB, Maine, promoted to SSgt/E-5. With hiring by the Nashua, Fire Department pending, I went to inactive status. Employed by the Nashua Fire Dept. six years.

Moved to Northern New York, 1984. Joined Saranac Vol. F.D., enlisted in New York Army National Guard as a medic (91A-20), rank of Sgt. Attended Combat Medical Specialist School in 1986, was an honor graduate. After two years went inactive, until Operation Desert Shield. Enlisted in Army Reserve, 310th Field Hospital, Malone, New York.

Active duty in 1991 - 92, attended college for LPN (91C-20) school, on President's list. 1993, as a lieutenant, and department chaplain, was injured battling house fire, reinjured working as a nurse. Due to injuries, left the reserves permanently, and was classified as permanently, totally disabled. Had numerous operations.

My wife and I live in Winter Haven, Florida. We have three grown children, one a 1Lt in the USAF, and four grandchildren.

Cornelius Kimball "C.K." McCusker

Former Captain, USAF, earned a B.S. degree in 1968 from The University of Arizona. Attended OCS in 1969 earning a commission as a 2nd Lt. Attended Navigator Training and Nav Bomb Training at Mather AFB, California, and was assigned to the 325th Bomb Squadron, 92nd Bomb Wing, Fairchild AFB, Washington. TDY to Guam in Apr '72 as one of the first three B52 Gs deployed to SEA combat ops. Flew Linebacker II and earned six Air Medals flying from Andersen, AFB, Guam, and U-Tapao RTAFB in Thailand.

McCusker left the USAF in 1974 and joined General Electric as a member of an 18-month internal management program which led to senior management positions at GE in the following 12 years. In 1986, he was recruited to start NEC Home Electronics (U.S.) which grew very rapidly and was a major competitor in the top tier of the consumer electronics industry challenging Sony and Mitsubishi.

He began an entrepreneurial career in 1993 and founded a number of successful start-ups related to his expertise in consumer products marketing. He was recruited back to NEC-Packard Bell in 2000 to manage their consumer computer business, but grew tired of the weekly commute to California after two years, and returned to Scottsdale where he went on to manage the U.S. operations of numerous other companies.

Based on the insistence of a Navy captain, and good friend, he began part-time writing a novel in 2005, centered on Linebacker II entitled *Yankee Air Pirates*. It was published in 2010 by Deeds Publishing in Atlanta. The book met with rave reviews including one from General Richard Myers, (USAF, Ret.), former Chairman, JCS, who said "*Yankee Air Pirates* is the *Catch-22* and *M*A*S*H* of the Vietnam War…"

McCusker is now a novelist, President of an LLC, and business consultant residing in Scottsdale, Arizona. Married for 42 years, he is blessed with three adult children, and three grandchildren.

Bruno Melton

Lt Colonel, USAF (Ret.) is a native of San Antonio, Texas. After graduation from MacArthur High School in 1983, he attended Trinity University and graduated with a Bachelor of Computer Science degree in 1987.

Upon graduating from pilot training in 1991at Vance Air Force Base, Oklahoma, he was selected to fly B-52s at Barksdale AFB, Louisiana. After a five-year tour, Colonel Melton went to teach students in the T-37 at Laughlin AFB, Texas.

Next, he was selected to teach pilots to become instructors at Randolph where he lead two flights as flight commander separated by a TDY as a NATO liaison officer in Italy during Operation NOBIL ANVIL.

His next assignment was back to Barksdale where he upgraded to instructor and selected as the 96th Bomb Squadron Standardization and Evaluation Flight Commander.

After 9/11, he flew in two deployments flying 21 combat sorties for Operation ENDURING FREEDOM which included Operation ANACONDA and Operation IRAQI FREEDOM where he flew the initial Shock and Awe airstrike dropping 12 Joint Direct Attack Munitions and eight cruise missiles.

Colonel Melton was sent on a "hardship" tour with his family to Hickam AFB, Hawaii. He served as Chief, Operations Plans and Information Operations planning over 220 bomber missions throughout the region.

In 2007, he was assigned back to Randolph Air Force Base and the 559th Flying Training Squadron as a T-6 Pilot Instructor Training instructor.

Colonel Melton is a command pilot with over 5,200 hours of military flying time. He currently works as a T-6 simulator instructor at Randolph AFB, Texas.

AWARDS AND DECORATIONS: Meritorious Service Medal with three oak leaf clusters, Air Medal, Aerial Achievement Medal with oak leaf cluster, Air Force Commendation Medal with oak leaf cluster, Joint Service Achievement Medal, Combat Readiness Medal, Kosovo Campaign Medal, and the Humanitarian Service Medal.

Gelvin Nicely

Chief MSgt, USAF (Ret.), was sworn in to the Air Force in June 1970 by his father, an active-duty Air Force colonel. As a geodetic computations specialist at the 1st Combat Evaluation Group, Barksdale AFB, Louisiana, he completed tours of duty at the ground-directed radar bombing detachments (call sign "Milky") at Hue Phu Bai and Son Tra (Monkey Mountain) Bases, Republic of Vietnam. In 1972, he was assigned to the 1st Geodetic Survey Squadron, F. E. Warren AFB, Wyoming, where he performed astronomic position, elevation, and gravity anomaly surveys and computations.

Nicely retrained in 1974 and served as a B-52G gunner and flight instructor, a B-52H instructor and flight examiner, and wing gunnery simulator instructor at Ellsworth AFB, South Dakota (1974-82). He was reassigned to Grand Forks AFB, North Dakota, where he served as B-52G training flight instructor gunner, wing flight scheduler and flight examiner (1982-85).

Nicely was selected for reassignment to Headquarters SAC, Offutt AFB, Nebraska, in 1985. As an exercise manager, he directed SAC participation in RED FLAG, GREEN FLAG, and MAPLE FLAG exercises, and managed SAC's nuclear bomb simulators as well as all live and simulated conventional weapons used for exercises and training. In 1987 he was reassigned as SAC acquisition manager for programs to meet operational needs for lethal and non-lethal aircraft defensive systems and worldwide connectivity improvements.

In 1992 he retrained again and was assigned as Superintendent, Physiological Training Center, Holloman AFB, New Mexico. In 1995 he was reassigned as Enlisted Manager of the 9th Physiological Support Squadron at Beale AFB, California, conducting support activities for global U-2 reconnaissance, including flight line operations, pressure suit maintenance, and specialized aircrew training. After over 30 years of service, Chief Nicely retired and was sworn out of the Air Force in July 2000 by his father, a retired Air Force colonel.

Vincent "Vince" H. Osborne

Colonel, USAF (Ret.) received a BA in Education from the University of Northern Colorado and a commission through AFROTC. Attended Navigator Training and Electronic Warfare Officer Training at Mather AFB, California. Following B-52 CCTS training in B-52F models at Castle AFB, California in 1968, he was assigned to the 486th Bomb Squadron at March AFB, California flying B-52E and F-models. He was assigned to temporary duty (TDY) at Westover AFB to become qualified in B-52 D-models. He was stationed TDY to Guam in 1968 for a short tour flying Operation Arc Light missions, and in 1970 to Guam, Thailand, and Okinawa for more Arc Light.

Returning to March AFB he became an instructor and a member of a Standardization/Evaluation crew. He was transferred to Carswell AFB in late 1971 and flew as an instructor of those preparing to go to fly Arc Light missions. Later Operation Bullet Shot sent many more B-52 crews to the Southeast Asia (SEA) arena. Vince volunteered for a Bullet Shot deployment during which he flew two Linebacker II missions. In 1973 he served his last deployment to SEA. During these deployments he flew 123 missions over SEA.

He served at SAC headquarters in the Aircraft Test Division and completed his Master of Arts in Business Management, the Air Force Electronic Warfare School in Sacramento, Supreme Headquarters

Allied Powers in Europe, Belgium in the Long Term Defense Program Office, Unified Space Command in Colorado Springs, at the Pentagon in Office of the Secretary of the Air Force for Acquisition for two years then two more years in the Office of the Undersecretary of Defense for Acquisition completing 26 years of service. He was awarded the Distinguished Flying Cross, Defense Meritorious Service Medal with one Oak Leaf Cluster (OLC), Meritorious Service Medal /w three OLC, Air Medal with one silver and three bronze OLC among other decorations and awards.

Michael W Radke

Master Sergeant, USAF (Ret.) Enlisted in the Air Force in 1957. After Basic training, assigned to Defensive Fire Control School at Lowery AFB. Upon completion he was assigned to Castle AFB. In 1960 he returned to Lowery AFB as a DFCS Instructor. In 1962 he returned to Castle AFB, and in 1965 started Gunnery school. Assigned to Glasgow AFB. TDY to Arc Light in 1966, '67, '68, and participated in the support of the USS Pueblo (Port Bow). In May 1968 Glasgow closed. Transferred to H-models at K I Sawyer AFB. TDY back to Arc Light in 1970. Participated in the RAF Blue Steel Trophy competition, and won in April 1971.

Retired in 1978, and employed as an Electrical Supervisor at Fort McCoy, Wisconsin. Retired after 20 years. At age 66 moved to Hubertus ,Wisconsin, to pursue a degree at UW, Anthropology / Archaeology.

Ray Schalk

Former Captain, USAF, was born and raised in Louisville, Kentucky, and attended the University of Louisville, earning a BA degree in Psychology. He received his commission through the AFROTC Program. Ray entered pilot training in Dec. '63 in the class of 64-D at Craig AFB in Selma, AL. He learned to fly in T-37s and T-Birds. He remembers the flight training as being fun and a breeze. Academics, on the other hand, were tougher. B, C, and D classes got a disproportionate number of Academy guys. "Bad News" – they set a high curve for a psyche major. "Good News" – they were always there, willing to tutor/coach the less scientific students – even into the wee hours. SAC was powerful in those years, and as Ray remembers, a disproportionate share of UPT assignments were – 47s, with a few 52s, and 135s and even less of others – GO FIGURE! He was pleased when he got one of the B-52 slots.

After the normal cold war B-52 training route – Nuclear Weapons, Castle AFB, and Stead – he was assigned to the 824th Bomb Squadron at Turner AFB, Georgia and served his first TDY Arc Light tour in mid-1966. Upon their return, Ray moved with the squadron to the 736th Bomb Squadron at Columbus AFB in Mississippi. In June, he departed for his second Arc Light tour. Then Ray departed the Air Force, having fulfilled his five-year military commitment. He logged over 2,000 hours and had 111 combat missions. His decorations include the DFC and the Air Medal with four OLC. Ray believes that his Air Force career was more than 'rewarding' and one of the most character building experiences of his life.

Ray went to work for Procter and Gamble as a sales manager. He continued to fly C-97s, after joining the 105th MILALFT Squadron, ANG in Nashville, Tennessee. P&G promotions took the Schalks to various U.S. locations. Ray retired from P&G in 2,000 after 32 years of service. Ray and Alice have three children and six grandchildren. They split their time between their winter residence, in Clermont, Florida, and their mountain home in western North Carolina.

Kenneth Schmidt

Major, USAF (Ret.) entered the USAF through Air Force OTS in January, 1974, and received his commission in April, 1974. He was assigned to Mather AFB, California, where he completed Undergraduate Navigator Training (UNT) flying in the T-29 and T-43 aircraft earning his navigator wings and completing the Navigator/Bombardier Training (NBT) course in April, 1975. He was then assigned to the 9th Bomb Squadron, 7th Bomb Wing, at Carswell AFB, Texas, serving in the B-52D until 1980. His next assignment was the 23rd Bomb Squadron, 5th Bomb Wing at Minot AFB, North Dakota, flying in the B-52H on combat crews as a radar navigator and instructor, and later served as a 5th Bomb Wing Bomber Aircrew Scheduler.

In 1983, Ken was transferred to Castle AFB, California, as a 4018th CCTS Academic Instructor in the B-52G. He later served as Chief, Air Weapons Branch. In 1987, he moved to Offutt AFB, Nebraska, and flew with the 2nd Airborne Command and Control Squadron (2nd ACCS) as an Operations Plans Officer onboard the EC-

135C "Looking Glass". Ken was later assigned to HQ/SAC DOO and held various staff jobs in the DO community until 1992 when he was assigned to the 55th Strategic Recon Wing, working in the Wing Inspector's Office until his retirement in 1993. He accumulated 3,651 flying hours: 155 hours in trainers; 2,681 in the B-52D, G, and H; and 811 hours (100 operational missions) in the EC-135C.

Ken completed his Master's Degree from Texas Christian University, and Squadron Officer's School, Air Command and Staff College, and Air War College.

Ken was married to Nancy (Schuler) in 2006, and they now reside in Bellevue, Nebraska. They have four children: Ryan Schuler of Sioux Falls, SD, Chris Schuler of Fargo, ND, Staci (Schmidt) Ewing of Mesa, AZ, and Stephanie (Schmidt) Fuhrman of Elkhorn, NE; and six grandchildren.

Dan Staley

Major, USAF (Ret.) After graduating from The Citadel, The Military College of South Carolina in 1971 Dan attended Undergraduate Pilot Training at Laughlin AFB in Del Rio, Texas. He completed Combat Crew Training at Castle AFB, California and was posted to Grand Forks AFB, North Dakota as a copilot. As a 1st Lieutenant he upgraded to pilot and became the aircraft commander for an all lieutenant crew. He was briefly transferred to Carswell AFB, Texas, and later to Dyess AFB. Dan was promoted to Flight Commander follow a SAC reorganization and then quickly moved to the Squadron Training Officer role. He finished his tour at Dyess as the Chief of the Bomber Stan/Eval Division.

Dan transferred to Barksdale AFB, LA as a Command Evaluator in the 1st Combat Evaluation Group and later was promoted to Chief of Bomber Branch for 1CEVG. During this timeframe he led the transformation of 1CEVG from strictly Flight Evaluation to Tactics Training & Evaluation. Dan retired in March of 1991. He participated in NATO Reforger 85-88, Team Spirit, SAC's Global Shield, and Desert Storm. Dan is a graduate of Squadron Officer School, Air Command and Staff College, National Security Management School in Washington DC, and has a Master of Arts degree from Louisiana Tech University. During his 20 year career as a member of the Strategic Air Command Dan amassed over 5,000 hours in the B-52D, F, G, and H-models.

Robert B. Stewart Jr

Lt Colonel, USAF (Ret.) Enlisted AF August 1951. Assigned as technical instructor at Lowry AFB, Colorado, in Officer Personnel and Career Guidance School. Discharged 1952 to accept a reserve commission. Completed navigation training at Ellington AFB, Texas, and Electronic Warfare training at Keesler AFB, Missippi, in December, 1953.

Assigned to 67th Tactical Reconnaissance Wing, K-14, Kimpo, Korea, in January, 1954. Assigned two months as EW advisor to First Marine Air Wing, Korea. Reassigned to 548th Recon Tech Sq, Yokota AB, Japan, in 1954. Assigned to Hq FEAF, Tokyo, Japan, 1954, as mission planner for RB-50G reconnaissance aircraft.

Assigned to Loring AFB, Maine, to await training in B-52. 1956. Admitted to Chelsea Navy Hospital for rheumatoid arthritis treatment and later surgery at Walter Reed Hospital.

Assigned to Keesler AFB, Mississippi, in 1957 with duties as ECM classroom and flight instructor, intelligence library director and briefing officer for the USAF ECM familiarization team. Accepted a regular commission as a captain with no break in service from reserve active duty status.

Assigned to Glasgow AFB, Montana, 1962-1967 as a B-52D EW instructor and evaluator. Completed a six-month Arc Light assignment to Andersen AFB, Guam.

Assigned to Headquarters, 15AF, March AFB, California, as staff EW officer, Air Training Branch, DOTA, 1967-1971. Refused HqSAC assignment to allow daughter to complete school. Then was immediately assigned to Dyess AFB, Texas, 1971-1975 to 96 Bomb Wing. Assigned temporary duty to U-Tapao AB, Thailand, as staff officer and mission briefer. Later augmented the flight training staff at Carswell AFB, Texas.

Retired September, 1975 with 24 years of active duty. Logged 5,654 hours of flight time, mostly in B-52D airrcraft with 630 hours of combat time.

George R. Thatcher, Ed.D

Major, USAF (Ret.) George Thatcher joined the Air Force in 1951, soon after graduating from his hometown high school in West Orange, New Jersey. He spent five years as an administrative specialist and training NCO, then attended and graduated from Air Force OCS Class 1957-B. After two years as an Air Defense Weapons Director, he was accepted into pilot training and graduated in Class 1960-H.

George chose to fly the "heavies" and was assigned as a B-52 copilot at Turner AFB, Georgia, where he spent over six years, upgrading to aircraft commander along the way. He flew many airborne alert missions, especially during the Cuban missile crisis, and had eight months of combat flying out of Anderson AFB, Guam. When Turner AFB closed in 1967 he was assigned as aircraft commander and instructor pilot to Plattsburgh AFB, New York.

In 1969 he volunteered for another tour of duty in Southeast Asia, this time flying the EB-66 "Destroyer" at Takhli RTAFB, Thailand. Following that experience, he spent his final two years of active duty as a Command and Control Officer at Blytheville AFB, Arkansas.

Since retiring from the Air Force in 1972, George has been a real estate broker, mortgage loan officer, middle/high school Spanish teacher, prison educator, and university professor of education. He is retired from Texas Tech University and presently serves as Adjunct Professor of Workforce Education for Southern Illinois University's off-campus degree program. His major passion is writing/collecting aviation poetry and short stories, and he engages in a variety of fitness activities. He gives up the game of golf frequently.

George is married to Jean Schwisow-Thatcher, his Beauteous Wife and Princess for Life, whom he credits with teaching him everything he knows about the education profession.

Tommy Towery

Major, USAF (Ret.) earned a B.S. degree in Journalism in 1968 from Memphis State University where he also earned a commission as a 2nd Lt. through the AFROTC program. Attended Navigator Training and Electronic Warfare Officer Training at Mather AFB, CA. Following B-52 CCTS training in B-52F models at Castle AFB, CA he was assigned to the 20th Bomb Squadron at Carswell AFB, TX flying B-52C and B-52D models. He was deployed for six months to Guam as part of "Operation Bullet Shot" and was assigned to 8th Air Force Bomber Operations as an Arc Light mission planner. On his next deployment he worked as an 43rd Bomb Wing Arc Light planner and flew on B-52 combat missions as a staff officer and helped plan Linebacker II missions. In 1976 his crew received the Mathis Trophy, awarded to the top bomber unit based on combined results in bombing and navigation in the SAC Bomb-Nav Competition.

He logged over 1,600 hours in B-52s and over 5,000 hours total flight time. His decorations include the Meritorious Service Medal with oak leaf cluster, the Air Medal with eight oak leaf clusters, the Armed Forces Expeditionary Service Medal, the Republic of Vietnam Campaign Medal and the South Vietnam Cross of Gallantry with Palm. He retired from the University of Memphis in 2008 and lives in Memphis, Tennessee, with his wife Sue who graciously allows him to devote many hours to his writing hobby. Tommy has written three non-military books, *"A Million Tomorrows –Memories of the Class of '64," "While Our Hearts Were Young," and "Goodbye to Bob."* He is a writer and editor of six other books in the "*We Were Crewdogs*" series and a contributing editor and publisher of a revision of *"Linebacker II – A View from the Rock."* He is a member of The American Author's Association and The Military Writers Society of America.

Dave Volker

Former Captain, USAF, has been a practicing safety professional for over 35 years providing consultative, training, and supervisory services to educational, industrial, and military clients across the North American continent. His expert, standardized approach to the process of machine safety evaluation, design, fabrication, and installation is based on almost two decades of guiding a staff of field safety representatives for a major third party administrator of insurance.

In the past 10 years he has attained national machine safety expert status working successfully with such diverse and technically challenging clients as General Electric Aerospace, Amazon.com, Pratt and Whitney Jet Engines, Lockheed Martin Undersea Systems, United Airlines, Goodrich Aerospace, Rolls-Royce Engines, Valero Energy, and the National Aeronautics and Space Administration (NASA) at Johnson Space Center, Houston, Texas. In this capacity he advised LRM's clients and employees in design, fabrication, and installation of a wide range of innovative machine guards and controls. For the past seven years, he has been designated the primary machine safety instructor by the Minnesota Department of Education for recurring certification classes.

Dave graduated from the University of North Dakota with a triple major in natural sciences; biology, geology and chemistry. He took that knowledge to the USAF and served as a B-52 pilot for seven years, flying 42 combat missions and accumulating over 3,000 flying hours. While he served in this capacity he was also educated in and performed the duties of a Nuclear Weapons Safety Officer. In this role, Dave learned and applied the most effective techniques of accident prevention, employee training, and documentation. The importance of professionally performing his military duties was reinforced by the unimaginable consequences of failure to excel.

Dave lives in Oakdale, MN. You may reach him at MrSafetyToo@Gmail.com or call him at 651-249-1980.

Walter W. Want

Former Captain, USAF, graduated thru ROTC from New Mexico State University in 1964 and entered pilot training at Williams AFB, Arizona. He was assigned to the 319th Bomb Wing at Grand Forks AFB, ND, flying the B-52H. He was a copilot and was awarded the trophy for Busy Figure II, Most Stable Flight Condition during Boeing's B-52 service test. Awarded the 1968 Junior Officer of the year and just prior to upgrade to AC status he was removed from flight status and entered into a program that led him to NKP, Thailand. As a Forward Air Controller with the 23 TASS he flew the only "push me-pull me" aircraft in the inventory, the mighty Cessna O-2-A. He flew day and night along with a special assignment to SOG. As a FAC he survived two bombings by Arc Lights and was happy to return to civilian life.

His awards include several Air Medals and two DFC. He makes his home in Oregon, just up the road from the Evergreen Air and Space Museum.

Dave Wellman

Former Captain, USAF, Born in Lafayette, Indiana, Dave attended Purdue University where he received a BSEE degree and an Air Force Reserve Commission in 1959. After marriage to Juliana

Kalberer, also of Lafayette, he worked at Chrysler in Detroit. He entered the Air Force in December 1959 and was in one of the last contract primary classes flying T-34's and prop driven T-28's at Spence AB, Georgia. He then flew T-33's at Laredo AFB, Texas and received his Wings in January 1961. After B-47 training at McConnell AFB, Kansas and Little Rock AFB, Arkansas, he was stationed at Lockbourne AFB, Ohio. In 1964, Lockbourne's B-47's were phased out and he went to Castle for B-52 training before going to Mather AFB, California. He had two tours at Andersen and during this time upgraded to aircraft commander.

He left the AF in 1967 to become an engineer for Boeing in Wichita, Kansas. He worked on the test and integration of SRAM A Missiles at White Sands Missile Range, New Mexico. He returned to Mather as a tech rep to an ATC Navigator Training Squadron and was assigned to NBT to assist in developing B-52 and FB-111 training courses for SRAM A flight crew operation. He and his family eventually moved to Seattle where he worked on the test and integration of the ALCM and on developmental testing of military and commercial aviation avionic systems. In retirement he is a Docent at Seattle's Museum of Flight where he works with children's education programs and is part of the speaker's bureau giving talks to community groups and schools about the B-52 and its involvement in the Vietnam War. He and his wife now live on Mercer Island, Washington where they enjoy life with each other and their three children and nine grandchildren.

Personal Notes

Linebacker II – A View from the Rock

Introduction by Maj. Tommy Towery (Ret.)

Brig. General James R. McCarthy
Lt. Colonel George B. Allison
General Editor, Colonel Robert B. Rayfield

This book is a narrative drawn from the era of the Southeast Asian conflict, detailing a unique event in that lengthy struggle. The event was called LINEBACKER II, and for the first time in contemporary warfare, heavy jet bombers were employed in their designed role to conduct extended strategic operations against the warmaking capacity of a hostile nation. This was the massive bombing of North Vietnam at the end of 1972.

This monograph tells part of the story of Strategic Air Command's participation in LINEBACKER II. In so doing, it addresses the efforts of a complex mixture of Air Force and sister service operations, with all services working in concert towards a common goal. Rather than develop a complete chronology or blow-by-blow account, which are matters of record in other works, the campaign is pursued more from the personal perspective.

Linebacker II: A View from the Rock was first published in 1976 by the Airpower Research Institute of the Air War College, located at Maxwell Air Force Base, Alabama.

Now available from www.wewerecrewdogs.com

Flying from the Black Hole

The B-52 Navigator-Bombardiers of Vietnam

By ROBERT O. HARDER

U.S. AIR FORCE NAVIGATORS AND BOMBARDIERS have long labored under the shadow of pilots, their contributions misunderstood or simply unknown to the public. This was especially the case with the B-52 non-pilot officer aircrews in the Vietnam War. Yet without them it would have been impossible to execute nuclear war strike plans or fly conventional bombing sorties. Here, one of their own reveals who these men were and what they did down in the "Black Hole." It is the only work to detail the B-52 air war in Vietnam from the perspective of a navigator-bombardier.

THE BOOK BEGINS WITH the war's climactic 1972 Hanoi Christmas bombing, an operation so poorly planned that it nearly became a disaster of epic proportions. The author then offers a background history of the development of bombing techniques and the evolution of bomber aircraft, focusing on the Vietnam-era B-52. The final chapters return to the eleven-day "Christmas War" over Hanoi and Haiphong for an insider's view of that defining battle, described by the author as the last massed, heavy bomber raid the world will ever see. Harder concludes his history with a discussion of the B-52 and its capabilities in the twenty-first century.

<u>Available at local bookstores or Amazon.com.</u>

Yankee Air Pirates

The Catch 22 and M.A.S.H. of the Vietnam War

By C.K. McCUSKER

IN THE WINTER OF 1972 President Nixon ordered Operation Linebacker II, the most massive aerial bombing campaign since the end of World War II. In this successful effort to bring the North Vietnamese back to the bargaining table and release our prisoners of war, B-52s were sent from Thailand and from the US island territory of Guam to attack Hanoi and Haiphong, North Vietnam, cities protected by more surface-to-air missiles than any other metropolitan complex outside of Moscow, Russia. Between December 18 and 29, in over 1,000 sorties, the B-52's dropped nearly 20,000,000 pounds of bombs, creating enough destruction and noise to convince the communist government of North Vietnam to sue for peace.

YANKEE AIR PIRATES tells the story of the air campaign from the eyes of the 20-something-year-old men who flew their massive Stratofortress into extreme threat environments, and mostly lived to tell about it. Yankee Air Pirates will introduce you to colorful characters: Spike, the best bombardier in SAC... at least in his own mind; Bulldog, Spike's navigator, close friend, and by his own design, an ongoing personal irritant to Spike; D-Kid, a 24-year-old navigator who became the first BUFF crew member to be shot down and captured; and a host of others who flew hard, played harder, and succeeded in getting our prisoners of war released.

Order now from www.yankeeairpirates.com or amazon.com